ANATOMY & PHYSIOLOGY

for beauty and complementary therapies

Workbook

Ruth Hull

thewriteidea

Published by The Write Idea Ltd
8 Station Court
Station Road
Great Shelford
Cambridge
CB22 5NE

01223 847765

First published April 2010
ISBN 978-0-9559011-2-6

Set in 10/12.5 Sabon.
Printed by Scotprint, Haddington

Contents

The editorial team

Author

Ruth Hull is a freelance writer who specialises in natural health. Born and educated in Zimbabwe, she completed a degree in Philosophy and Literature before studying and practising complementary therapies in London. She now lives in South Africa where she lectures in complementary therapies at the renowned Jill Farquharson College in Durban. Ruth is married with two children.

General editor

Greta Couldridge has worked within the industry for many years, starting her career in a salon environment, before moving to teaching posts in further education colleges where she progressed to a management role. Throughout this time she worked as an external verifier and examiner before accepting a position with an awarding body developing qualifications and assessment materials. Greta is continually developing her skills, technical knowledge and keeping abreast of changes within the industry by attending workshops, courses and conferences.

Contributor

Hazel Godwin is a lecturer in Anatomy & Physiology in the School of Hairdressing, Beauty and Related Therapies at Blackpool and the Fylde College in Lancashire. She has taught beauty and complementary therapy students there for the past 11 years. She is also a qualified registered nurse and midwife working most recently in intensive care nursing. Hazel has found the new text book to be of immense value and was delighted to be asked to contribute to the multiple choice questions in this workbook.

Reviewer

Manjit Gill trained in complementary therapies at Stephenson College, Leicestershire in 2001. In 2003 she completed her teacher training course and then taught massage as well as working in a salon and as a freelance therapist. She completed her degree in complementary therapies at Derby University in 2006 and then became a full time tutor in complementary therapies. She has taught Level 3 Anatomy and Physiology for many years and has also worked for City and Guilds revising test papers as well as producing material for their smartscreen website.

Publisher

Andy Wilson has worked in the publishing industry for 23 years, setting up The Write Idea in 1991. Most recently he has worked with one of the leading international therapy awarding bodies, helping to set up their publishing division and co-producing a number of successful text books.

Introduction

Welcome to our anatomy & physiology workbook. The purpose of this book is to help you succeed in the difficult task of learning the anatomy and physiology of the human body. To do this, it has been written in two parts.

Part One

Firstly, the body has been broken down into systems and each system is represented by a chapter in which:

- You are firstly guided through making your own revision and self-study notes. This section suggests exercises that will help you to both understand and learn the facts you need to know. You may come across some study tips or suggestions for group work. **Please note, answers are not provided for this section.**
- You then work through a series of questions and exercises that will test your knowledge of the body system you are studying. These questions include a section on the pathology of the system, a short vocabulary test and multiple choice questions. **Answers to all these exercises are provided at the end of the book.**

Part Two

Once you have covered all the chapters in Part One and you feel confident that you have a sound understanding of each system and a good knowledge of the facts, you can move on to Part Two of the workbook. This section is made up of 14 multiple choice papers, each one covering the entire anatomy and physiology syllabus. Pretend you are taking an exam and, under exam conditions, work through a paper. When you mark it, make a note of the questions you got wrong and your weak areas. Before moving on to the next multiple choice paper spend some time revising your weak areas.

Memory Techniques

Some of the revision and study guide exercises ask you to draw mind maps orspider diagrams, write cue cards or create mnemonics. This is how to do it. **Remember – nobody else needs to see any of these so don't worry if you are not an artist!**

Mind maps and spider diagrams

Mind maps are simple diagrams that help you link thoughts, ideas or information. They are fun to create and it is while creating them that you begin to understand and enjoy what you are learning. It is also easier to remember images that you have personally created than to remember text that has been written by somebody else.

How to create a mind map:

- Take the text that you need to learn or understand and underline the key ideas. Now convert these written ideas into images.
- On a blank piece of paper draw the central idea and surrounding it draw the other ideas.
- Link them up with arrows or lines.

What is important when creating a map is to use words, colours, pictures or symbols that are personal to YOU. Use things that you will remember.

A practical example of a mind map

This example has been taken from Chapter 3 where you need to learn about eccrine glands. The 'e' is for eccrine, the hands and feet are reminders of the location, the thermometer is a reminder of their function, and the sweaty face is a reminder of their excretion.

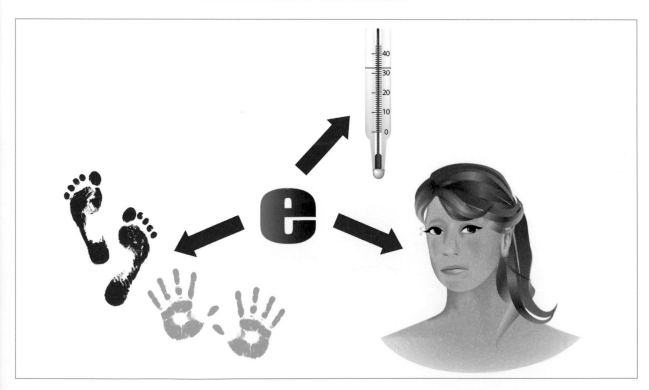

Cue cards and sticky notes

Cue cards and sticky notes are short, simple revision notes that contain only the facts that need to be memorised. When you create and use cue cards remember to:

- Write important definitions, names, dates etc on little cards that you can carry with you everywhere you go. When stuck in a queue or waiting for a bus you can have a quick look at one or two of them.
- Write important points on sticky notes and stick them on your fridge, toilet door, telephone, bathroom mirror or anything you see on a daily basis.
- The important thing is to keep them short and simple. Use colour and/or symbols.

A practical example of a cue card

The word 'caudal' means away from the head or below.

Caudal	Away from the head, below.
Front of card	Back of card

Mnemonics

Mnemonics are words, sentences or verses created to help you remember information. When you create a mnemonic remember to:

- write a list of the facts you need to recall.
- underline the first letter of each fact.
- create a rhyme, phrase or word that you will easily remember using these letters.

The more personal, fun and amusing these phrases are the easier you will remember them.

Practical example of a mnemonic

The epidermis of the skin is composed of five layers. These are the stratums:

> **B**asale, **S**pinosum, **G**ranulosum, **L**ucidum, **C**orneum.
>
> **B**ad **s**kin **g**rafts **l**ook **c**omical

1 Before you begin

Revision/Self-study notes

Before you can study anatomy you need to have a good knowledge of anatomical terminology. Unfortunately, the only way to learn this terminology is by memorising it. To help you do this, make cue cards of the following terms and their definitions:

Anatomy	Caudal	Lumbar	Gluteal
Physiology	Ventral	Coxal	Crural
Pathology	Cutaneous	Pelvic	Femoral
Anatomical position	Cephalic	Inguinal	Patellar
Anterior	Facial	Pubic	Popliteal
Posterior	Frontal	Perineal	Sural
Dorsal	Ophthalmic/Orbital	Acromial	Tarsal
Superior	Otic	Scapular	Pedal
Inferior	Buccal	Axillary	Calcaneal
Medial	Nasal	Brachial	Plantar
Lateral	Occipital	Cubital	Parietal
Proximal	Cervical	Antecubital	Visceral
Distal	Thoracic	Olecranal	Sagittal plane
Superficial	Costal	Antebrachial	Frontal/coronal plane
Deep	Pericardial	Carpal	
Peripheral	Mammary	Manual	Transverse plane/ cross-section
Cephalad	Abdominal	Palmar/Metacarpal	Oblique plane
Cranial	Umbilical	Digital/Phalangeal	

Studytip

Write the definitions in your own words. It is always easier to remember your own words than someone else's.

Study groups

To help you revise your teminology play the game **30 seconds**. Here's how to do it:

30 seconds
Preparation:

1. Make or buy some small cards. On each card write down five different anatomical terms. The more cards you have the longer the game will be.
2. Get a clock or stop watch so that you will be able to time 30 seconds per round.
3. Divide yourselves into two teams (teams A and B).

To play:

1. One member from team A will stand up, take a card and quickly read the five terms to themself.
2. That person then has to describe each term to their team mates without saying the term written on the card.
3. The team mates need to guess and call out the terms written on the card. They will get one point for each term they identify correctly.
4. Each team has only 30 seconds in which to go through the terms on the card.
5. At the end of the round write down how many points that team has obtained.
6. It is then team B's turn.
7. At the end of the game, when all the cards have been used up, add up the points each team has gained. The team with the most points wins!

Exercises

1. This figure shows a man standing in the anatomical position.
   Label the arrows with the directional terms listed on the right.

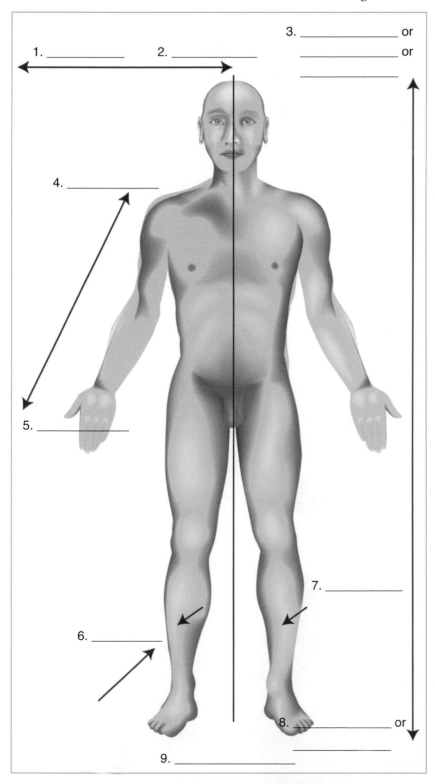

3. _____ or

   _____ or

   _____

1. _____ 2. _____

4. _____

5. _____

6. _____

7. _____

8. _____ or

   _____

9. _____

Median line
Superior
Inferior
Medial
Lateral
Proximal
Distal
Superficial
Deep
Cephalad
Cranial
Caudal

2. Using directional terms, fill in the blanks:

    a. The heart is _____ to the stomach.

    b. The diaphragm is _____ to the lungs.

    c. The heart is _____ to the lungs.

    d. The lungs are _____ to the heart.

    e. The knee is _____ to the ankle.

    f. The ankle is _____ to the knee.

    g. The hands and feet are at the _____ of the body.

3. Match the anatomical term to the region it describes:

Anatomical term	Region described
1. Buccal	a. Hollow behind the knee
2. Cephalic	b. Back
3. Dorsal	c. Elbow or forearm
4. Phalangeal	d. Head
5. Cubital	e. Heart
6. Inguinal	f. Skin
7. Popliteal	g. Fingers or toes
8. Pericardial	h. Cheek
9. Sural	i. Calf
10. Cutaneous	j. Groin

Answers:

1. ___ 2. ___ 3. ___ 4. ___ 5. ___ 6. ___ 7. ___ 8. ___ 9. ___ 10. ___

4. The figure below shows the cavities of the body. Label the cavities and place the organs listed into the cavities in which they belong.

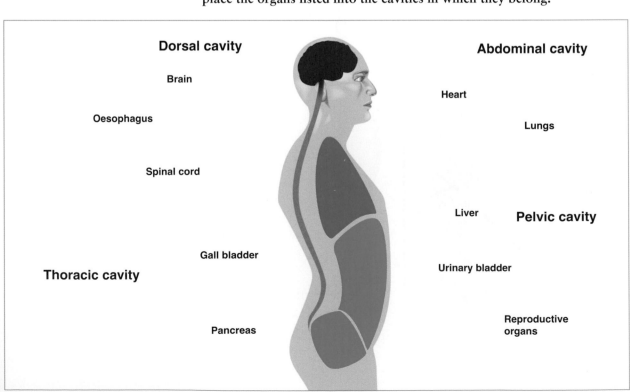

Dorsal cavity

Brain

Oesophagus

Spinal cord

Thoracic cavity

Gall bladder

Pancreas

Abdominal cavity

Heart

Lungs

Liver **Pelvic cavity**

Urinary bladder

Reproductive organs

5. The figure below shows the planes of the body. Label them.

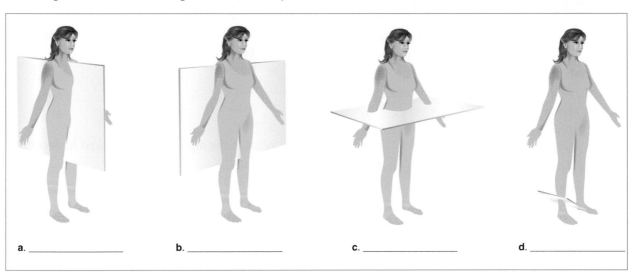

a. _____   b. _____   c. _____   d. _____

Vocabulary test

Complete the table below.

WORD	DEFINITION
Anatomy	a. _____
Cephalad	b. _____
Cross-section	c. _____
Distal	d. _____
Midline	e. _____
Parietal	f. _____
Pathology	g. _____
Physiology	h. _____
Ventral	i. _____
Visceral	j. _____

Multiple choice questions

1. **Which of the following phrases is correct?**
   a. Anatomy is the study of the functions of the body.
   b. Anatomy is the study of the diseases of the body.
   c. Anatomy is the study of the structure of the body.
   d. Anatomy is the study of the pathology of the body.

2. **If a man is standing in the anatomical position he will be standing:**
   a. Facing forward with his arms hanging by his side and his palms facing forward.
   b. Facing forward with his arms hanging by his side and his palms facing backward.
   c. Facing backward with his arms hanging by his side and his palms facing forward.
   d. Facing backward with his arms hanging by his side and his palms facing backward.

3. **Another term for the midline is:**
   a. The middle line
   b. The median line
   c. The medial line
   d. The periphery.

4. **Which of the following phrases is correct?**
   a. The term cephalad means towards the feet.
   b. The term deep means towards the surface of the body.
   c. The term proximal means farther from the point of attachment of a limb.
   d. The term dorsal means at the back of the body.

5. **Which of the following structures are all located in the abdominal cavity?**
   a. Stomach, urinary bladder, reproductive organs.
   b. Liver, spleen, gallbladder.
   c. Pancreas, small intestine, bronchi.
   d. Lungs, heart, diaphragm.

6. **The term 'calcaneal' describes which region of the body?**
   a. Thigh
   b. Calf
   c. Ankle
   d. Heel.

7. **What is the anatomical term used to describe the hips?**
   a. Coxal
   b. Inguinal
   c. Costal
   d. Umbilical.

8. **The term 'parietal' relates to:**
   a. The internal organs of the body.
   b. The external organs of the body.
   c. The inner walls of a body cavity.
   d. The external walls of a body cavity.

9. **What type of plane divides the body vertically into posterior and anterior portions?**
   a. Coronal
   b. Transverse
   c. Sagittal
   d. Oblique.

10. **What is the collective name of the regions that the abdominopelvic cavity can be divided into?**
    a. Fifths
    b. Quadrants
    c. Quarters
    d. Thirds.

2 Organisation Of The Body

Revision/Self-study notes

Levels of structural and chemical organisation of the body

When you first start studying anatomy it helps to have a basic understanding of how the body is organised and the chemicals it is made up of.

1. List the six levels of structural organisation of the body.

_____   _____

_____   _____

_____   _____

2. Name the four major elements that make up 96% of the body's mass:

_____   _____

_____   _____

3. Create a mnemonic for the nine lesser elements in the body. These are:
- Calcium
- Phosphorous
- Potassium
- Sulphur
- Sodium
- Chlorine
- Magnesium
- Iodine
- Iron

Refer to page vii of the introduction to see how to create a mnemonic.

Mnemonic _____

4. Describe each of the following compounds and their role in the body:

a. Water _____

_____

_____

_____

b. Carbohydrates _____

_____

_____

_____

c. Lipids (fats) _____

_____

_____

_____

d. Proteins _____

_____

_____

_____

e. Nucleic acids _____

_____

_____

_____

f. Adenosine triphosphate _____

_____

_____

_____

Cellular organisation of the body

Cells are the basic structural and functional units of the body and although there are many different types of cell, they generally have the same basic make-up. They are surrounded by a plasma membrane which regulates the movement of all substances into and out of the cell.

5. Describe the structure of the plasma membrane in your own words: _____

_____

_____

6. Describe the function of the plasma membrane in your own words: _____

_____

_____

7. Substances can be transported across the plasma membrane by a number of different processes. Describe each of the following processes in no more than 10 words. The first one has been done for you:

a. Simple diffusion: _from high concentration to low concentration_ _____

b. Osmosis: _____

c. Facilitated diffusion: _____

d. Active transport: _____

e. Vesicular transport: _____

8. Imagine the cell is a factory and list all the cell structures as parts of that factory. A similar example to this is given on page 31 of *Anatomy & Physiology for Beauty and Complementary Therapies*, where the cell has been likened to a bakery (e.g. the plasma membrane is the walls and doors of the building).

The cell as a factory:

- Plasma membrane _____

- Cytoplasm _____

- Nucleus_____

- Mitochondria_____

- Ribosomes _____

- Endoplasmic reticulum_____

- Golgi complex _____

- Lysosomes _____

- Peroxisomes_____

9. In the box below, draw a spider diagram or mind map showing the lifecycle of a cell.

Tissue level of organisation of the body

Tissues are composed of groups of similar cells all working together to perform a common function.

10. Complete the following table on tissues in the body. The first key phrase has been done for you.

TYPES OF TISSUE IN THE BODY			
Name	Description	Function	Example
Epithelial			
Key phrase to remember	Epithelial tissue covers and lines.		
Connective			
Key phrase to remember			
Muscle			
Key phrase to remember			
Nervous			
Key phrase to remember			

11. **There are so many different types of tissue in the body that it is often difficult to remember them all. For each tissue listed below write down no more than five words that will remind you of it. The first one has been done for you.**

Epithelial tissue
- Simple squamous (pavement) epithelium: <u>single, flat, lining</u>

- Simple cuboidal epithelium _____

- Simple columnar epithelium _____

- Ciliated simple columnar epithelium _____

- Stratified squamous epithelium _____

- Transitional epithelium_____

Connective tissue
- Areolar tissue_____

- Adipose tissue _____

- Lymphoid (reticular) tissue _____

- Elastic connective (yellow elastic) tissue_____

- Dense regular connective (white fibrous) tissue _____

- Osseous tissue (bone) _____

- Hyaline cartilage _____

- Fibrocartilage_____

- Elastic cartilage _____

- Vascular tissue (blood) _____

Muscle tissue
- Skeletal muscle tissue _____

- Cardiac muscle tissue _____

- Smooth (visceral) muscle tissue _____

Nervous tissue
- _____

Membranes
- Mucous membranes _____

- Serous membranes _____

- Connective tissue (synovial) membranes _____

Exercises

1. Word search – Can you find the 13 elements present in the body?

D	M	A	G	N	E	S	I	U	M	O	P
G	A	C	A	L	C	I	U	M	Y	S	O
R	S	H	F	C	F	T	Q	C	T	C	T
A	H	Y	J	I	U	U	E	A	L	H	A
I	L	D	R	H	G	I	H	R	N	E	S
G	Y	R	F	V	O	R	F	B	B	I	S
P	H	O	S	P	H	O	R	O	U	S	I
G	I	G	O	K	P	N	U	N	D	U	U
K	G	E	D	H	F	F	Y	M	W	L	M
V	F	N	I	T	R	O	G	E	N	P	F
I	S	Q	U	F	Q	X	T	T	D	H	C
K	I	R	M	Q	D	Y	W	Q	D	U	F
O	T	R	E	E	I	G	W	F	O	R	K
K	I	O	D	I	N	E	Y	G	O	T	P
C	H	L	O	R	I	N	E	M	U	G	H

2. Fill in the missing words:

a. _____compounds do not contain the element carbon. An example of such a compound is _____ . This compound functions in maintaining the _____ of the body by absorbing and giving off large amounts of heat; acting as a _____ around organs, bones, ligaments, tendons and the gastrointestinal tract; cushioning and protecting organs such as the _____; and acting as a _____ into which many different substances can dissolve and be transported around the body.

b. Organic compounds contain the element _____ . Carbohydrates, _____, _____, _____ _____ and _____ _____ are all examples of important organic compounds present in the body.

c. Carbohydrates contain the elements _____, _____ and oxygen and provide _____ for the body.

d. Lipids are commonly called _____ and contain the elements

_____, _____ and _____.

e. Proteins contain the elements _____, _____,

_____, _____ and sometimes _____.

3. Link each organic compound with its role in the body and write your answers in the spaces given below.

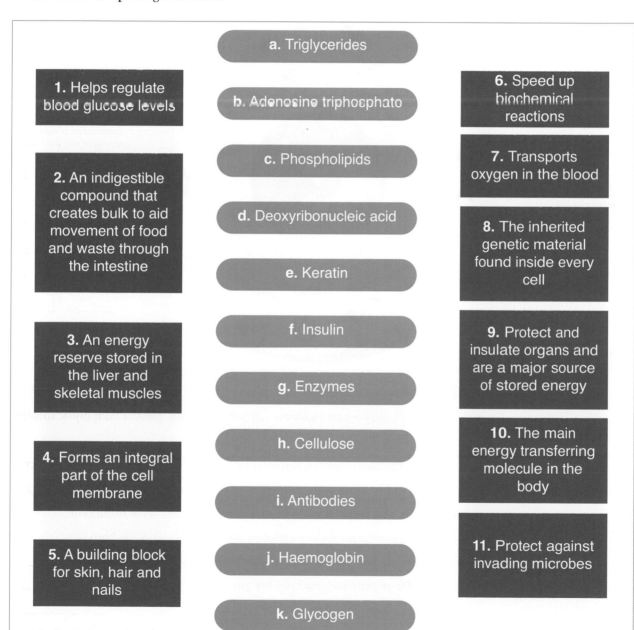

1. Helps regulate blood glucose levels

2. An indigestible compound that creates bulk to aid movement of food and waste through the intestine

3. An energy reserve stored in the liver and skeletal muscles

4. Forms an integral part of the cell membrane

5. A building block for skin, hair and nails

a. Triglycerides

b. Adenosine triphosphate

c. Phospholipids

d. Deoxyribonucleic acid

e. Keratin

f. Insulin

g. Enzymes

h. Cellulose

i. Antibodies

j. Haemoglobin

k. Glycogen

6. Speed up biochemical reactions

7. Transports oxygen in the blood

8. The inherited genetic material found inside every cell

9. Protect and insulate organs and are a major source of stored energy

10. The main energy transferring molecule in the body

11. Protect against invading microbes

Answers:

1. ___ 2. ___ 3. ___ 4. ___ 5. ___ 6. ___ 7. ___ 8. ___ 9. ___ 10. ___ 11. ___

4. **The figure below shows a generalised animal cell. Complete the labelling of this diagram by filling in the blank spaces.**

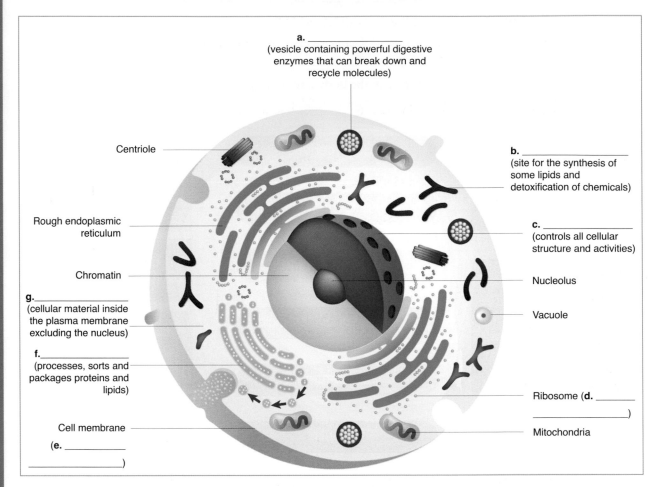

a. _____
(vesicle containing powerful digestive enzymes that can break down and recycle molecules)

Centriole

Rough endoplasmic reticulum

Chromatin

g._____
(cellular material inside the plasma membrane excluding the nucleus)

f._____
(processes, sorts and packages proteins and lipids)

Cell membrane

(**e.** _____
_____)

b. _____
(site for the synthesis of some lipids and detoxification of chemicals)

c. _____
(controls all cellular structure and activities)

Nucleolus

Vacuole

Ribosome (**d.** _____
_____)

Mitochondria

5. **Are the following statements true or false? Write T or F in the spaces provided.**

a. In vesicular transport particles are carried across the plasma membrane in sacs. _____

b. Osmosis is the movement of water from an area of high solute concentration to an area of low solute concentration. _____

c. Simple diffusion is the movement of water only from areas of high concentration to areas of low concentration. _____

d. Somatic cell division takes place when the body needs to replace dead and injured cells or produce new cells for growth. _____

e. Human cells have 23 chromosomes. _____

f. In somatic cell division a single diploid parent cell duplicates itself into two identical daughter cells through a process called meiosis. _____

g. Through the process of mitosis a mother cell divides into two identical daughter cells. _____

h. Cytokinesis occurs during mitosis and is the period in which chromosomes go to the opposite ends of the cell and the mitotic spindle breaks down and disappears. _____

i. Histology is the study of the different types of cells in the body. _____

j. There are four types of tissue in the body: epithelial, connective, muscle and nervous.

_____

6. **Complete the following table on the different tissue types found in the body.**

TYPES OF TISSUE IN THE BODY			
Tissue type	**Description**	**Function**	**Example**
a.	A loose, soft and semifluid type of connective tissue containing collagen, elastic and reticular fibres.	Strength, elasticity and support.	Surrounds body organs.
Transitional epithelium	**b.**	Stretching and distension.	Bladder.
Fibrocartilage	Strongest type of cartilage consisting of cells scattered between bundles of collagen fibres.	Support and strength.	**c.**
Simple cuboidal epithelium	Single layer of cube-shaped cells.	**d.**	Forms ducts of many glands.
e.	A single layer of rectangular cells containing cilia.	Moving fluid or particles along a passageway.	Fallopian tubes.
Adipose tissue	An areolar tissue containing many fat cells.	**f.**	Subcutaneous layer of the skin.
g.	Consists of many freely branching elastic fibres, few cells and little matrix. Has a yellowish colour.	Stretching and strength.	Walls of arteries.
Cardiac muscle tissue	A muscle tissue consisting of branched, striated fibres not under conscious control.	Pumps blood.	**h.**
i.	Wet membranes lining body cavities that open to the exterior.	Lubrication, movement and protection.	Respiratory tract.
Connective tissue membranes/ synovial membranes	Composed of areolar connective tissue with elastic fibres and fat. They secrete synovial fluid.	**j.**	Line cavities of freely movable joints such as the shoulder joint.

Vocabulary test

Complete the table below.

WORD	DEFINITION
Phagocytosis	a. _____
Enzyme	b. _____
Homeostasis	c. _____
Diploid	d. _____
Vesicle	e. _____
Metabolism	f. _____
Somatic cell	g. _____
Gene	h. _____
Solvent	i. _____
Kinetic energy	j. _____

Multiple choice questions

1. The smallest unit of matter is a:
   a. Compound
   b. Chemical
   c. Atom
   d. Element.

2. The four major elements of the body are:
   a. Sulphur, molybdenum, oxygen and nitrogen.
   b. Hydrogen, oxygen, nitrogen and copper.
   c. Oxygen, nitrogen, hydrogen and zinc.
   d. Carbon, hydrogen, oxygen and nitrogen.

3. What group of organic compounds includes triglycerides and steroids?
   a. Carbohydrates
   b. Lipids
   c. Proteins
   d. Nucleic acids.

4. Which of the following is the main energy-transferring molecule in the body?
   a. ATP
   b. ADP
   c. DNA
   d. RNA.

5. Free radicals are:
   a. Molecules that are vital for the proper functioning of the body.
   b. Substances that combat or neutralise unstable cells.
   c. Molecules that are highly unstable and reactive and that damage cells.
   d. Substances that play a role in the growth and development of the body.

6. The movement of water across a selectively permeable membrane from an area of high water concentration to an area of low water concentration is called:
   a. Osmosis
   b. Simple diffusion
   c. Active transport
   d. Vesicular transport.

7. In a cell, proteins are synthesised in the:
   a. Mitochondria
   b. Lysosomes
   c. Centrosomes
   d. Ribosomes.

8. In a cell, ATP is generated in the:
   a. Mitochondria
   b. Lysosomes
   c. Centrosomes
   d. Ribosomes.

9. How many chromosomes do you inherit from your father?
   a. 13
   b. 23
   c. 36
   d. 46.

10. The time in which a cell grows and prepares itself for division through DNA replication is called:
    a. Mitosis
    b. Meiosis
    c. Interphase
    d. Cytokinesis.

11. Which of the following has the stages of mitosis in their correct chronological sequence?
    a. Interphase, prophase, metaphase, anaphase, telophase, interphase.
    b. Anaphase, prophase, telophase, interphase, metaphase, anaphase.
    c. Telophase, anaphase, prophase, metaphase, interphase, telophase.
    d. Metaphase, interphase, prophase, telophase, anaphase, metaphase.

12. What type of tissue usually lines hollow organs, cavities and ducts?
    a. Epithelial
    b. Connective
    c. Muscle
    d. Nervous.

13. **What type of tissue protects and supports the body and also binds organs together and stores energy reserves?**
    a. Epithelial
    b. Connective
    c. Muscle
    d. Nervous.

14. **The words squamous, cuboidal, columnar and transitional all describe which type of tissue cells?**
    a. Epithelial
    b. Connective
    c. Muscle
    d. Nervous.

15. **The main function of lymphoid (reticular) tissue is:**
    a. Allows for distension.
    b. Energy reserve.
    c. Support framework.
    d. Secretion and absorption.

3 The Skin, Hair and Nails

Revision/Self-study notes

The skin

The skin is a cutaneous membrane made of two distinct layers, the epidermis and the dermis. When studying the skin it is important to know the difference between these two layers.

1. Complete the table below to highlight the differences between the epidermis and dermis.

DESCRIPTION	EPIDERMIS	DERMIS
Where in the skin is this layer found? Is it superficial or deep?		
What type of tissue is this layer composed of?		
What are the main functions of this layer?		
What types of cells does this layer contain?		
How many sub-layers is each layer composed of?		
Write a few words or a short phrase to remind you of this layer.		

Studytip

You may be the type of person who remembers images more easily than words. If you are, then use symbols or images instead of words. For example, when learning the stratum basale/germinativum in the following exercise on page 28, you could draw a seed beginning to germinate. Don't be afraid to have fun – remember, the more personal, colourful, amusing and fun an image or phrase is the easier you will remember it.

2. The epidermis is composed of five different layers. In the space below draw the different layers of the epidermis and next to each write a short phrase or draw an image or symbol that will remind you of this layer.

3. The dermis is composed of only two layers, the papillary and reticular layers. Write a short phrase to remind you of each layer:

- Papillary layer _____

- Reticular layer _____

4. Although it is not part of the skin, you need to learn about the subcutaneous layer of tissue beneath the skin. This tissue attaches the reticular layer of the dermis to the underlying organs. What type of tissue is this layer made up of?

   _____

5. There are five main skin types: normal/balanced, oily, dry, combination and sensitive. In the chart on the next page draw five different faces and on each face draw in the main characteristics of each skin type. Use lots of colours and your imagination and have fun!

Normal/balanced

Oily

Dry

Combination

Sensitive

Study groups
Bring your anatomy and physiology studies to life. Form a group and analyse each other's skin type. Remember to not just look at the skin but also feel it for greasiness, coarseness or dryness and ask questions about health and lifestyle. Once you have seen and felt the different skin types you won't forget them.

Hair (pili)

1. Humans have three different types of hair – lanugo, vellus and terminal.
Complete the table below:

HAIR TYPE	DESCRIPTION	WHERE IS IT FOUND?
Lanugo		
Vellus		
Terminal		

2. The figure below shows the basic structure of a hair. Label these areas:
follicle, shaft, root.

3. Using your own words, describe the life cycle of a hair.

_____

_____

_____

_____

_____

_____

Nails

When learning about the nails you might feel overwhelmed by all the different terms you need to learn. Unfortunately, there is no easy way to learn all these terms and you will just have to memorise them. **Make up some cue cards to help you remember the following terms:**

- Germinal matrix
- Nail bed
- Cuticle
- Eponychium
- Nail plate
- Free edge
- Nail groove
- Nail mantle
- Lunula
- Hyponychium
- Nail wall

Study groups

Play a game with the cue cards. One person stands up in front of the group and calls out either the term or the definition on the card. The members of the group then compete to call out the correct answer. Whoever calls out the correct answer first wins a point. At the end of the game the person with the most points wins!

Cutaneous glands

Glands are groups of specialised cells that can either secrete beneficial substances or excrete waste products from the body. The main glands associated with the skin are sebaceous and sudoriferous (sweat) glands.

1. **In your own words, write down a simple phrase to remind you of each type of gland:**

- Sebaceous glands _____

- Sudoriferous glands _____

2. Sudoriferous glands can be divided into eccrine and apocrine glands depending on their location, structure, secretion and function. Draw spider diagrams or mind maps to help you remember each of them.

Common pathologies of the skin, hair and nails

Learn both the clinical and common names of these diseases and disorders and you will notice that you already know many of them. For example, you know what nail biting is – now you just need to learn its clinical name: onychophagy.

As a therapist you will need to have a basic understanding of a wide variety of skin diseases and disorders. Write down a short phrase that will remind you of each of the following. The first one has been completed for you:

Abnormal growth disorders

- Keloids _overgrowth of scar tissue_ _____

- Psoriasis _____

- Seborrheic keratoses (senile warts) _____

- Verrucae filliformis (skin tags) _____

Allergies

- Contact dermatitis_____
- Eczema _____
- Urticaria (nettle rash or hives)_____

Bacterial infections of the skin

- Carbuncles _____
- Cellulitis _____
- Conjunctivitis_____
- Folliculitis_____
- Furuncles (boils) _____
- Hordeolum (stye) _____
- Impetigo _____

Cancers

- Basal cell carcinoma (rodent ulcers) _____
- Melanoma _____
- Squamous cell carcinoma (prickle-cell cancer) _____

Fungal infections of the skin

- Candidiasis (thrush or yeast infection)_____
- Tinea capitis (scalp ringworm) _____
- Tinea corporis (body ringworm)_____
- Tinea pedis (athlete's foot) _____

Infestations of the skin

- Pediculosis capitis (head lice)_____
- Pediculosis corporis (body lice)_____
- Pediculosis pubis (pubic lice) _____
- Scabies (itch mites) _____

Pigmentation disorders

- Albinism_____
- Chloasma _____
- Dilated capillaries _____

- Ephelides (freckles) _____
- Erythema _____
- Lentigines _____
- Vascular naevi _____
- Vitiligo _____
- Pressure sores (decubitus ulcers) _____

Sebaceous gland disorders

- Acne rosacea _____
- Acne vulgaris _____
- Comedones (blackheads) _____
- Milia _____
- Seborrhoea _____
- Steatomas (sebaceous cysts or wens) _____

Sudoriferous (sweat) gland disorders

- Anhidrosis _____
- Bromhidrosis (body odour) _____
- Hyperhidrosis _____
- Miliaria rubra (prickly heat) _____

Viral infections of the skin

- Herpes simplex (cold sore)_____
- Herpes zoster (shingles) _____
- Rubella (German measles) _____
- Rubeola (measles) _____
- Varicella (chickenpox) _____
- Warts/verrucae_____

Pathologies of the nails

- Agnail (hang nail) _____
- Anonychia _____
- Beau's lines_____
- Egg shell nails _____

- Koilonychias (spoon nail)_____
- Leuconychia (white nails or white spots) _____
- Longitudinal furrows _____
- Onychauxis (thick nails)_____
- Onychogryphosis (ram's horn nails) _____
- Onychocryptosis (ingrowing nail)_____
- Onycholysis _____
- Onychophagy (nail biting) _____
- Onychoptosis (nail shedding) _____
- Onychorrhexis (brittle nails) _____
- Paronychia (bacterial infection of cuticle) _____
- Pterygium (overgrowth of cuticle)_____
- Severely bruised nail _____
- Tinea ungium (ringworm of the nail) _____

Studytip

Anatomy and physiology is a living subject and when learning it always try to remember you are learning about your own body. You can bring the pathologies of the skin to life by finding people who suffer from these conditions, looking at them and chatting to the person about them. It shouldn't be difficult to find someone who bites their nails (onychophagy); someone who has eczema, dermatitis or psoriasis; or someone who suffers from cold sores (herpes simplex).

Exercises

1. List seven functions of the skin.

a. _____

b. _____

c. _____

d. _____

e. _____

f. _____

g. _____

2. The figure below shows the anatomy of the skin. Complete the labelling of this diagram by filling in the blank spaces.

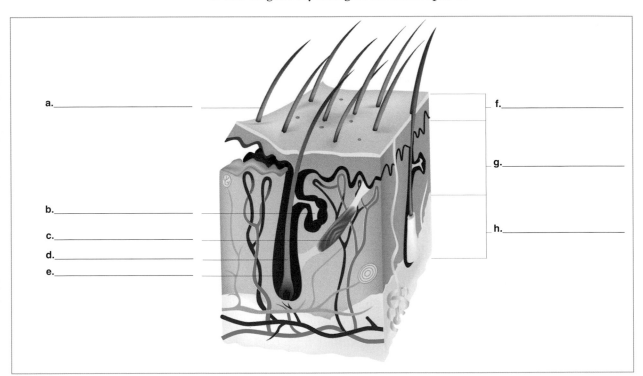

a._____

b._____

c._____

d._____

e._____

f._____

g._____

h._____

3. Are the following statements true or false? Write T or F in the spaces provided.

a. The skin is composed of two layers – the epidermis which contains the nerves, blood vessels, sweat glands and hair roots and the dermis which is continuously being worn away. _____

b. The epidermis is composed of keratinised stratified squamous epithelium. _____

c. The dermis is composed of areolar and adipose tissue. _____

d. The dermis is attached to a subcutaneous layer which anchors the skin to other organs of the body. _____

e. The subcutaneous layer is composed of connective tissue, collagen and elastic fibres. _____

4. The skin is made up of a number of different cell types. In the table below, match the cell type to its correct function and write the answers in the spaces provided.

CELL TYPE		FUNCTION	
1	Adipocytes	a	Produce a pigment that contributes to skin colour and absorbs ultraviolet light.
2	Fibroblasts	b	Produce a protein that waterproofs and protects the skin.
3	Keratinocytes	c	Engulf and destroy bacteria and cell debris through the process of phagocytosis.
4	Langerhans cells	d	Provide insulation.
5	Macrophages	e	Function in the sensation of touch.
6	Mast cells	f	Synthesise collagen, elastic and reticular fibres.
7	Melanocytes	g	Function in skin immunity.
8	Merkel cells	h	Produce histamine during inflammation.

Answers: 1. ____ 2. ____ 3. ____ 4. ____ 5. ____ 6. ____ 7. ____ 8. ____

5. The epidermis is composed of five layers. Put these layers into their correct order, working from the most superficial to the deepest layer.

1. Stratum spinosum 2. Stratum lucidum 3. Stratum basale 4. Stratum corneum 5. Stratum granulosum.

Correct order:_____

6. Fill in the missing words.

a. The _____ is the supportive layer beneath the epidermis and it can be divided into two layers, the _____ layer and the _____ layer.

b. The _____ layer is an undulating membrane composed of _____ _____ tissue and _____ fibres. It has nipple-shaped, fingerlike projections called _____ which project into the epidermis and contain loops of _____ so that diffusion of nutrients and oxygen can take place between the dermis and epidermis.

c. The _____ layer is composed of _____ _____ _____ tissue and contains both _____ and _____ fibres. This layer is the main support structure of the skin and it also contains _____ follicles, _____, oil _____, ducts of _____ glands and _____ tissue. The _____ layer is attached to the underlying organs by the _____ layer which contains _____ _____ tissue, _____ tissue and nerve endings sensitive to pressure.

7. Three clients come into your salon and each asks you what skin type they have. Next to each description below give the name of the skin type.

a. Susan's skin is flaky with dry patches and a fine texture. When you touch her skin it feels papery and you can see that she is ageing prematurely, especially around her eyes and mouth. Susan's skin type is:

_____.

b. Carl has good skin. It is smooth, even and blemish free. When you touch it, it is soft and firm and has good elasticity. Carl's skin type is:

_____.

c. Samantha has always had a problem skin. Most products make her skin red and itchy and she finds it is often dry and chafes easily. She also has dilated capillaries and when you touch her skin it feels unusually warm. Samantha's skin type is:

_____.

8. Complete the following crossword on the anatomy and physiology of the hair and nails.

Across
3. A function of both hair and nails (10).
7. Portion of the hair that penetrates into the dermis (4).
9. The soft hair covering a foetus while in the womb (6).
12. The transitional stage in the life cycle of a hair in which a fully grown hair detaches from the matrix (7).
13. An extension of the cuticle from the base of the nail fold (10).
16. The resting stage of the hair growth cycle (7).

Down
1. Another name for the nail mantle (4).
2. Sebaceous _____ glands secrete sebum which lubricates hair (3).
3. The arrector _____ is the muscle that pulls hairs into a vertical position (4).
4. _____ hairs prevent the inhalation of insects (7).
5. The superficial end of the hair that projects from the surface of the skin (5).
6. The crescent-shaped white area at the proximal end of the nail (6).
8. Type of hair found on the head (8).
10. The growing or active phase of the life cycle of a hair (6).
11. A nipple-shaped projection found in the hair bulb at the base of the follicle (7).
14. The Latin word for 'nail' (4).
15. The area lying directly beneath the nail plate (3).

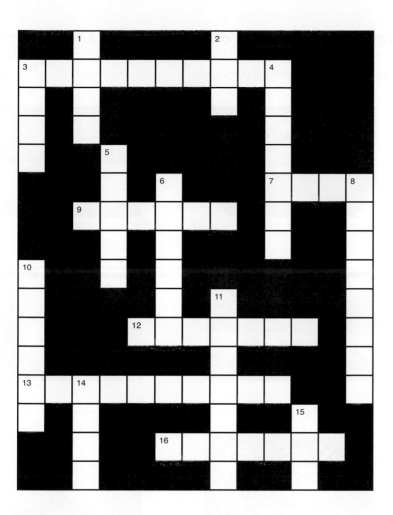

9. State whether the following are (1) sebaceous, (2) eccrine sudoriferous or (3) apocrine sudoriferous glands:

a. This type of gland excretes waste and helps regulate the body's temperature. _____

b. This type of gland is located in the axilla, pubic region and the areolae of the breasts. _____

c. This gland's secretory portion is located in the subcutaneous layer and its duct extends outwards through the skin, opening as a pore on the surface of the skin. _____

d. This is an oil gland that usually empties into hair follicles. _____

e. This type of gland secretes a lubricant that prevents excessive evaporation of moisture from the skin. _____

Common pathologies of the skin, hair and nails

10. Identify the following pathologies.

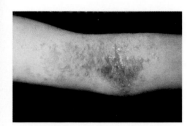

a. _____

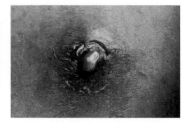

b. _____

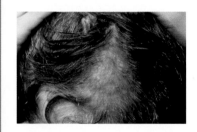

c. _____

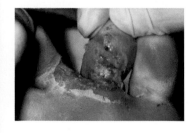

d. _____

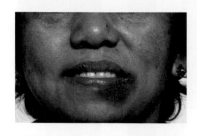

e. _____

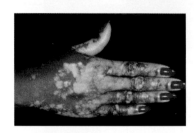

f. _____

g. _____

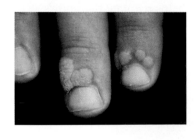

h. _____

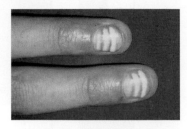

i. _____

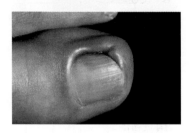

j. _____

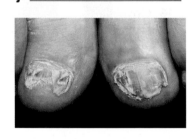

k. _____

11. In the table below, match the clinical name of a disease or disorder to its common name and write the answers in the spaces provided.

CLINICAL NAME		COMMON NAME	
1	Urticaria	a	Body ringworm
2	Hordeolum	b	Overgrowth of cuticle
3	Candidiasis	c	Hives
4	Pediculosis	d	Ingrowing nail
5	Tinea corporis	e	Measles
6	Ephelides	f	Stye
7	Erythema	g	Spoon nail
8	Comedones	h	Thrush
9	Herpes simplex	i	White spots on nail
10	Herpes zoster	j	Nail biting
11	Rubeola	k	Shingles
12	Varicella	l	Redness of skin
13	Koilonychias	m	Chickenpox
14	Leuconychia	n	Lice
15	Onychocryptosis	o	Cold sore
16	Onychophagy	p	Freckles
17	Pterygium	q	Blackheads

Answers

1. ___   2. ____   3. ____   4. ____   5. ____   6. ____   7. ____   8. ____   9. ____

10. ___   11. ____   12. ____   13. ____   14. ____   15. ____   16. ____   17. ____

Vocabulary test

Complete the table below.

DEFINITION	WORD
The transfer of infection from one person to another.	a. _____
The process through which the skin is shed.	b. _____
The system of the skin and its derivatives.	c. _____
The constriction of blood vessels.	d. _____
The dilation of blood vessels.	e. _____

Multiple choice questions

1. **Which of the following statements is correct?**
   a. Functions of the skin include excretion, secretion and movement.
   b. Functions of the skin include blood cell production, vitamin D synthesis and protection.
   c. Functions of the skin include sensation, absorption and heat regulation.
   d. Functions of the skin include protection, olfaction and sensation.

2. **Although the skin is waterproof, it can still absorb certain substances. These substances include:**
   a. Vitamins B and C
   b. Salts
   c. Water
   d. Fat soluble vitamins.

3. **Which of the following statements is true?**
   a. The skin is a serous membrane.
   b. The skin is a cutaneous membrane.
   c. The skin is a synovial membrane.
   d. The skin is a mucous membrane.

4. **What is the name of the thin, protective outer layer of the skin?**
   a. Dermis
   b. Epidermis
   c. Hypodermis
   d. Subcutaneous layer.

5. **What is the name of the deepest layer of the epidermis?**
   a. Stratum basale
   b. Stratum spinosum
   c. Stratum granulosum
   d. Stratum corneum.

6. **In which layer of the epidermis are all the cells dead and completely filled with keratin?**
   a. Stratum basale
   b. Stratum spinosum
   c. Stratum granulosum
   d. Stratum corneum.

7. **Through what process are skin cells constantly shed?**
   a. Phagocytosis
   b. Keratinisation
   c. Desquamation
   d. Vasoconstriction.

8. **What is the function of the reticular layer of the dermis?**
   a. It greatly increases the surface area of the dermis.
   b. It contains loops of capillaries so that diffusion of nutrients and oxygen can take place.
   c. It acts as the main support structure of the skin.
   d. It is the tissue that attaches the skin to the underlying organs.

9. **A skin that has an oily t-zone yet dry cheeks and neck can be described as:**
   a. Balanced
   b. Oily
   c. Dry
   d. Combination.

10. **Which of the following statements is correct?**
    a. Hairs are columns of living cells.
    b. Hairs are columns of reproducing cells.
    c. Hairs are columns of keratinised dead cells.
    d. Hairs are columns of dead cells covered in melanin.

11. **The area of a nail in which cell division and growth occurs is called the:**
    a. Germinal matrix
    b. Hyponychium
    c. Nail plate
    d. Lunula.

12. **The main glands associated with the skin are:**
    a. Endocrine and sebaceous glands.
    b. Endocrine glands only.
    c. Sebaceous and sudoriferous glands.
    d. Sudoriferous glands only.

13. **What type of gland secretes an oily substance called sebum?**
    a. Sebaceous
    b. Cutaneous
    c. Endocrine
    d. Sudoriferous.

14. **What type of infection is tinea corporis?**
    a. Bacterial
    b. Fungal
    c. Viral
    d. Yeast.

15. **What is the term used to describe inflammation or redness of the skin?**
    a. Chloasma
    b. Ephelides
    c. Erythema
    d. Lentigines.

4 The Skeletal System

Revision/Self-study notes

Bones

Bones play a vital role in both the structure and function of the body and they do far more than simply forming a skeleton to support it.

1. **List the functions of the skeletal system and next to each function write a sentence showing you understand the function.**

a. _____

_____

_____

b. _____

_____

_____

c. _____

_____

_____

d. _____

_____

_____

e. _____

_____

_____

f. _____

_____

_____

g. _____

_____

2. **To help you remember the different cells that make up bone tissue, create a short rhyme that demonstrates the following:**
- Osteoprogenitor cells are the initial stem cells which can develop into osteoblasts.
- Osteoblasts are the cells that form bones. They develop into osteocytes.
- Osteocytes are the mature bone cells that maintain the daily activities of bone tissue. They are the main cells found in bone tissue.
- Osteoclasts are cells found on the surface of bones that resorb or destroy bone tissue.

Rhyme:

_____

_____

_____

_____

3. **In the table below, list the differences between compact (dense) bone tissue and spongy (cancellous) bone tissue.**

CHARACTERISTIC	COMPACT (DENSE) BONE TISSUE	SPONGY (CANCELLOUS) BONE TISSUE
Basic description		
Basic structure		
Functions		
Location		

4. The bones of the body can be classified into the following types – long, short, flat, irregular and sesamoid. You will find these in the boxes below and on the following pages. Next to each one write its main features and functions. Then work your way through the skeleton, putting all the bones of the skeleton into their correct box. The first one, long bones, has been done for you.

Long bone

slightly curved to provide strength

greater length than width, contain a long shaft

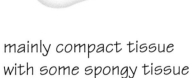

mainly compact tissue with some spongy tissue

Where in the body?
Femur, tibia, fibula, phalanges, humerus, ulna, radius

Short bones

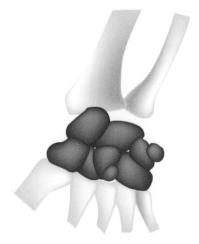

Flat bones

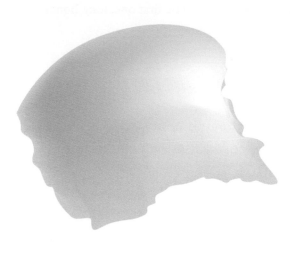

Irregular bones

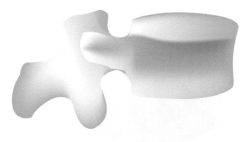

Sesamoid bones

5. **In the space provided, draw a long bone and label the following:**
   epiphysis, metaphysis, diaphysis, articular cartilage, spongy bone tissue,
   epiphyseal line, compact bone, periosteum, medullary cavity.
   Next to each label write the structure's function.

The structure of a long bone

The skeleton

1. You should now understand the structure, function and classification of bones. You now need to learn the skeleton. Colour in and label the skeleton below and as you work on each bone repeat its name to yourself a couple of times so that you can memorise it.

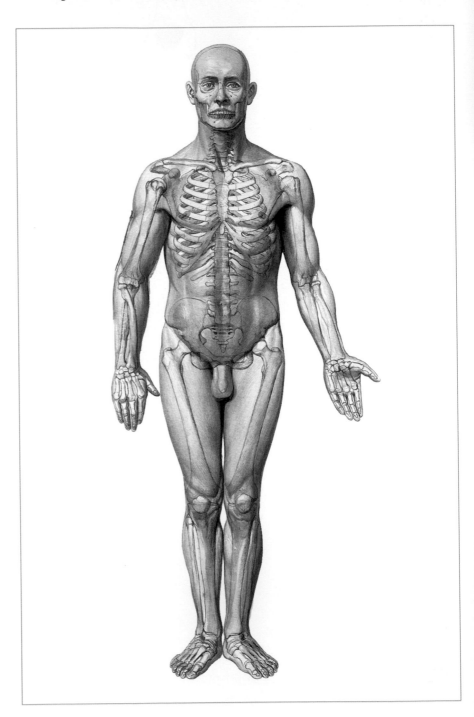

Studytip

Learn the names of the bones well because many muscles, blood vessels and nerves are named after the bones along which they run. Study hard now and it will make things easier for you later on!

2. To help you learn the bones of the cranium and face, blank keys and images of the skull have been provided below and on the next page. Choose your own colour for each bone, complete the keys provided and then colour in each bone.

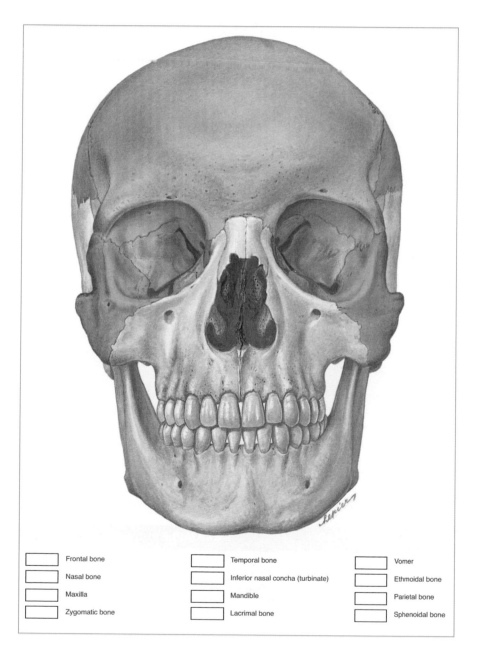

	Frontal bone		Temporal bone		Vomer
	Nasal bone		Inferior nasal concha (turbinate)		Ethmoidal bone
	Maxilla		Mandible		Parietal bone
	Zygomatic bone		Lacrimal bone		Sphenoidal bone

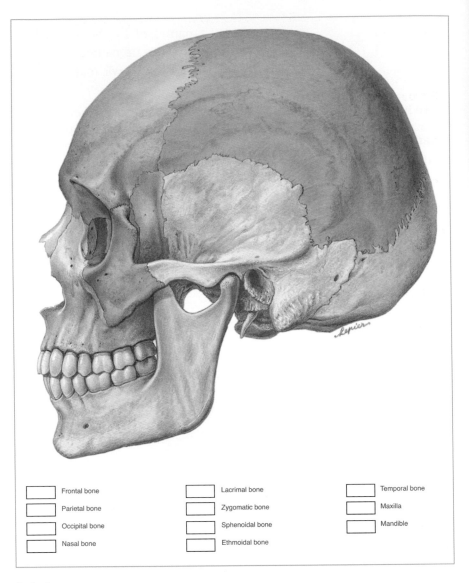

	Frontal bone		Lacrimal bone		Temporal bone
	Parietal bone		Zygomatic bone		Maxilla
	Occipital bone		Sphenoidal bone		Mandible
	Nasal bone		Ethmoidal bone		

Joints

Joints, or articulations, are the points of contact between bones and they function in not only holding bones together, but also in providing movement.

1. **Joints can be classified according to the degree of movement they permit. Complete the table below.**

CLASSIFICATION	MOVEMENT PERMITTED	EXAMPLE
Synarthroses		
Amphiarthroses		
Diarthroses		

2. Joints can also be classified according to their structure. Complete the
   table below.

CLASSIFICATION	STRUCTURE	EXAMPLE
Fibrous		
Cartilaginous		
Synovial		

3. Synovial joints permit a wide range of movements, all of which are listed
   below. Next to each movement write a few words or a short phrase to
   remind you of it.

• Flexion _____

• Extension _____

• Hyperextension _____

• Abduction _____

• Adduction _____

• Circumduction_____

• Rotation _____

• Medial (internal) rotation _____

  _____

• Lateral (external) rotation _____

  _____

• Pronation _____

• Supination _____

• Inversion_____

• Eversion _____

• Dorsiflexion_____

• Plantar flexion _____

4. There are a number of different types of synovial joints whose names tend to describe them. In the spaces provided, draw a symbol or image that will remind you of each type of synovial joint. Next to the symbol list or draw the movements it permits and give an example of the joint.

Gliding (plane)	**Hinge**
Pivot	**Condyloid**
Saddle	**Ball and socket (spheroidal)**

Common pathologies of the skeletal system

In the spaces provided write a few words or a short phrase that will remind you of the disease or disorder listed. The first one has been done for you.

Pathologies of the skeletal system

- Osteoarthritis *wear and tear, degeneration of cartilage, spurs in joint*

- Rheumatoid arthritis

- Bunion

- Bursitis

- Dislocation (luxation)

- Gout

- Osteogenesis imperfecta (brittle bone disease)

- Osteomalacia / rickets

- Osteoporosis

- Paget's disease

- Sprain

- Torn cartilage

Fractures

- Simple or closed fracture

- Compound or open fracture

- Comminuted fracture

- Impacted fracture

- Complicated fracture

- Greenstick fracture

- Stress fracture

Pathologies of the spine

- Cervical spondylosis _____

- Kyphosis _____

- Lordosis_____

- Prolapsed intervertebral disc (PID) _____

- Scoliosis_____

- Spinal stenosis _____

- Spondylitis _____

- Temporo-mandibular joint disorder (TMJ) _____

- Whiplash injury _____

Disorders of the feet

- Clubfoot (talipes equinovarus)_____

- Hammer toes _____

- Heel spur_____

- Pes cavus (high arches) _____

- Pes planus (flat foot) _____

Exercises

1. Word search – Can you find six functions of the skeletal system?

D	K	E	I	L	N	J	H	B	I	F	M	M	G	I
A	S	M	A	G	E	K	X	J	D	W	F	O	U	G
W	E	I	I	U	J	F	D	X	B	C	M	D	L	L
A	E	N	E	R	G	Y	S	T	O	R	A	G	E	D
S	E	E	U	I	O	S	X	D	Q	Z	E	D	Z	E
I	H	R	D	O	O	U	A	S	G	E	O	L	G	S
K	D	A	F	P	K	P	C	P	F	F	L	K	Y	Z
G	F	L	P	J	N	P	Y	R	D	J	T	H	M	U
V	H	H	A	E	M	O	P	O	I	E	S	I	S	T
N	C	O	U	G	A	R	U	T	G	L	I	Y	S	J
O	C	M	G	E	S	T	J	E	E	F	Y	U	A	I
U	O	E	D	S	Q	S	M	C	J	S	O	I	T	T
L	U	O	O	G	E	A	O	T	Z	J	G	G	F	O
N	D	S	J	V	F	G	E	I	L	G	D	X	S	I
C	K	T	T	M	A	O	M	O	V	E	M	E	N	T
S	L	A	A	B	I	T	A	N	D	S	A	W	O	A
R	F	S	E	D	F	J	X	G	X	Z	E	L	T	D
E	H	I	H	S	E	N	I	T	S	A	D	I	Q	S
S	A	S	N	Q	G	E	Y	O	B	Q	C	Y	A	E

2. Fill in the missing words.

a. _____ is the study of bone. Bone tissue is also called

_____ tissue and it is a _____ tissue.

b. Bone tissue is composed of water, protein, fibres and _____ salts.
The fibres of bone tissue are made of _____ which give
bones their tensile strength. The salts of bone tissue are mainly
_____ carbonate and a crystallised compound called
_____ . These salts give bone its _____ .

3. There are two different types of bone tissue. What are they?

a. _____

b. _____

4. Bones can generally be classified into five different types. The figure below gives examples of each of these. In the spaces provided label each type of bone.

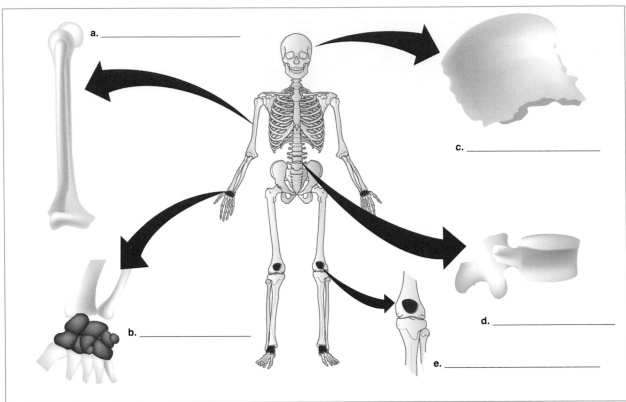

a. _____

b. _____

c. _____

d. _____

e. _____

5. The table below details the structure of a long bone. Match the term to its correct description and write the answers in the space provided.

TERM	DESCRIPTION
1 Diaphysis	a Membrane covering the space inside the diaphysis.
2 Epiphysis	b Area where bone shaft joins bone end.
3 Metaphysis	c Membrane covering the bone.
4 Epiphyseal plate	d Central shaft of bone.
5 Periosteum	e Space inside diaphysis.
6 Endosteum	f End of bone.
7 Medullary (marrow cavity)	g Layer of cartilage that allows bone to grow in length.

Answers: 1. ____ 2. ____ 3. ____ 4. ____ 5. ____ 6. ____ 7. ____

6. **Fill in the gaps.**

a. The _____ skeleton supports and protects the major

organs of the body while the _____ skeleton

forms the upper and lower extremities and their girdles.

7. **The following figures show the anterior and lateral views of the bones of the skull. Label the colour-coded keys below each diagram with the names of the bones.**

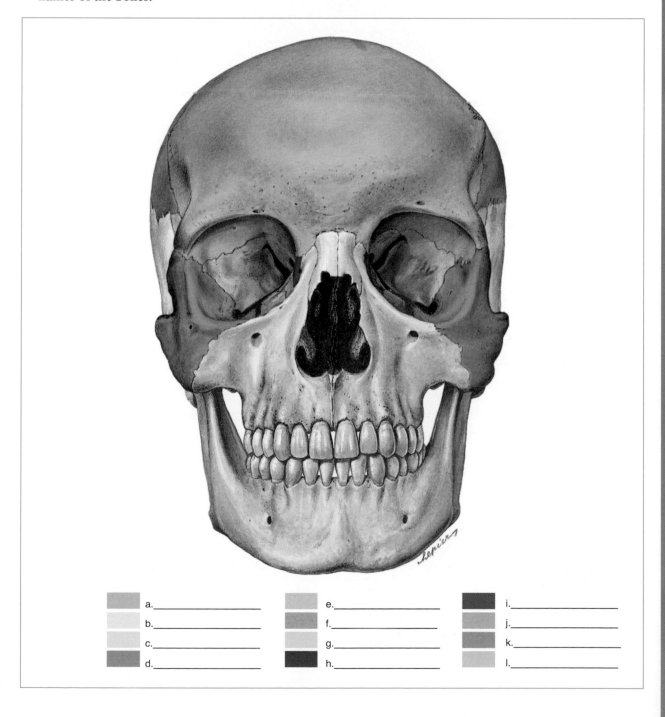

a._____   e._____   i._____

b._____   f._____   j._____

c._____   g._____   k._____

d._____   h._____   l._____

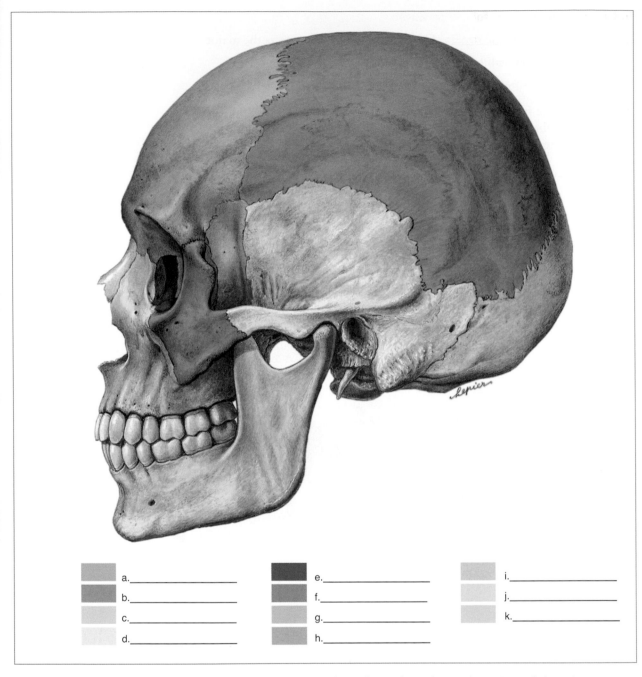

a._____

b._____

c._____

d._____

e._____

f._____

g._____

h._____

i._____

j._____

k._____

8. Give the correct number of vertebrae for each section of the spine.

a. Cervical vertebrae _____

b. Thoracic vertebrae_____

c. Lumbar vertebrae _____

d. Sacral vertebrae _____

e. Coccygeal vertebrae _____

9. **Give the names of the following bones:**

a. This bone supports the tongue and provides attachment for some of the
   muscles of the neck and pharynx _____ .

b. This is the breastbone _____ .

c. There are 12 pairs of these bones and they are located in the thorax
   _____ .

10. **The figure below shows the bones of the arm and shoulder girdle.
    Complete the labelling of these bones using the words on the right.**

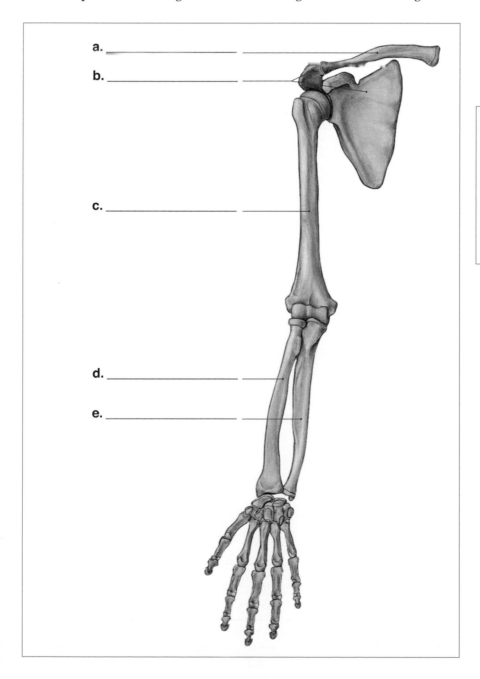

a. _____ _____

b. _____ _____

c. _____ _____

d. _____ _____

e. _____ _____

Radius

Clavicle

Ulna

Humerus

Scapula

11. The figure below shows the bones of the leg and pelvic girdle.
Complete the labelling of these bones using the words on the left.

Patella

Ilium

Tibia

Femur

Fibula

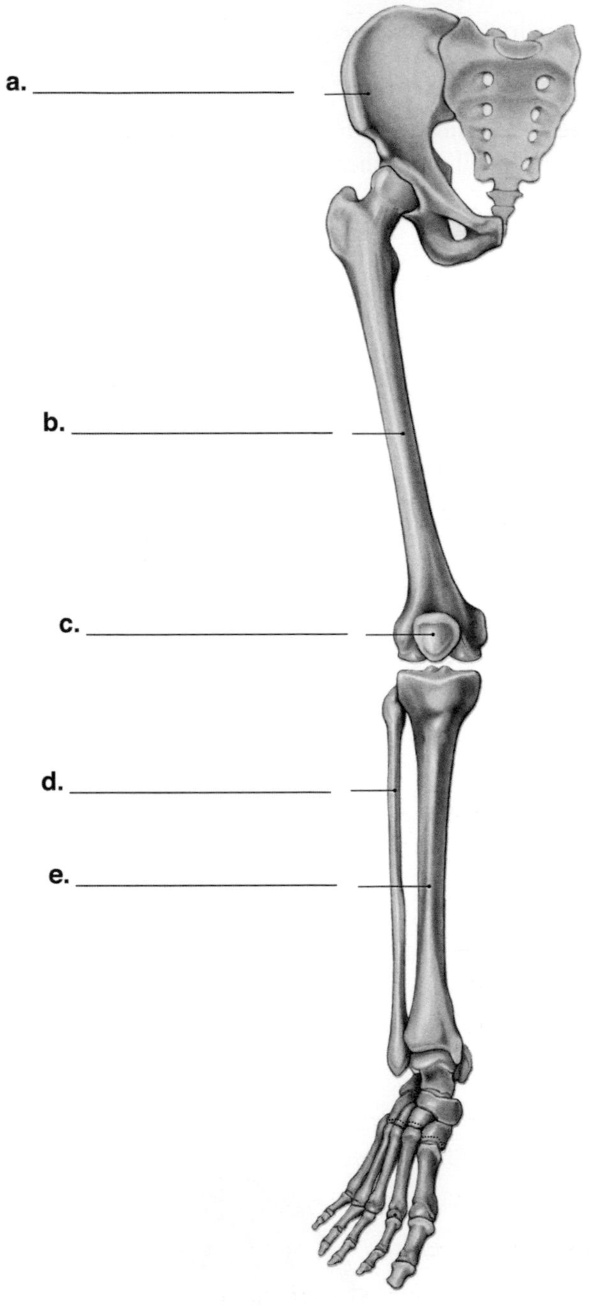

a. _____

b. _____

c. _____

d. _____

e. _____

12. **Listed below are the bones of the hands and feet. Connect the name of the bone to the area to which it belongs.**

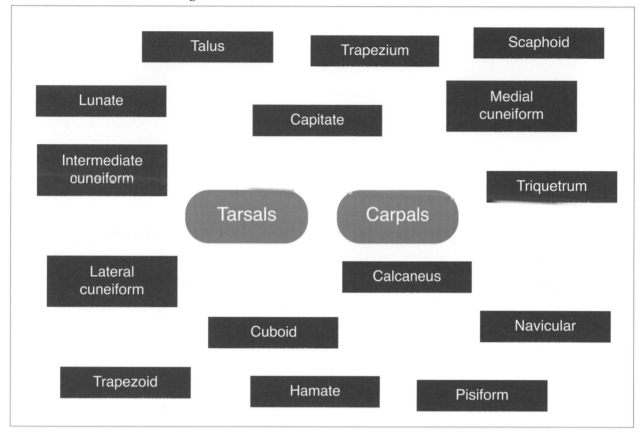

13. **Are the following statements true or false? Write T or F in the spaces provided.**

a. The bones of the feet can be arranged into four arches. These are the medial longitudinal arch, intermediate longitudinal arch, lateral longitudinal arch and the transverse arch. _____

b. A ligament is a band of connective tissue that attaches bones to bones. _____

c. Diarthrotic joints are immovable joints. _____

d. The coronal suture is an example of a synarthrotic joint. _____

e. In a synovial joint, the synovial cavity is the space that separates the articulating bones. _____

f. Synovial fluid is found between fibrous joints. _____

14. **In the gaps provided, give the correct term for each of the following movements.**

a. If you bring your chin down to your throat you are _____ your neck.

b. When doing star-jumps, if you swing your arms up and away from your body you are _____ them.

c. When turning your palms anteriorly, you are _____ them.

d. You _____ your arms when swinging them in a large circular movement as if you are warming up your shoulders.

15. Below are examples of different types of synovial joints. Give the name of each joint type.

a. Wrist joint _____

b. Elbow joint_____

c. Thumb joint_____

d. Joint between the axis and atlas _____

e. Sacro-iliac joint _____

f. Shoulder joint _____

Common pathologies of the skeletal system

16. Identify the following pathologies of the skeletal system.

a. The clinical term for a herniated or slipped disc is _____

_____ _____ or P.I.D.

b. This is an autoimmune disease in which the body attacks its own cartilage and joint lining. It is characterised by inflammation, swelling, pain and loss of function of joints and is thought to be hereditary. This disease is _____ _____ .

c. This is a fracture which only occurs in children and involves an incomplete break or crack in the bone. The bone then bends. This is a _____ fracture.

d. This disease usually affects men in middle-age or older and it involves the build-up of uric acid and its salts in the blood and joints. This disease is _____ .

e. An exaggerated thoracic curve of the spine is called _____ .

f. This disease is common in elderly and post-menopausal women and is a progressive disease in which bones lose their density and become brittle and prone to fractures. This disease is _____ .

g. An exaggerated lumbar curvature of the spine is called _____ .

h. Painful, rigid toes that are fixed in a contracted position and cannot be straightened are called _____ _____ .

i. The clinical term for high arches is _____ _____ .

j. A lateral curvature of the spine is called _____ .

Vocabulary test

Complete the table below.

WORD	DEFINITION
Tendon	a. _____
Ossification	b. _____
Ligament	c. _____
Haemopoiesis	d. _____
Articulation	e. _____

Multiple choice questions

1. **Which of the following are all functions of the skeletal system:**
   a. Movement, shape and vitamin D synthesis.
   b. Haemopoiesis, protection and support.
   c. Mineral homeostasis, heat generation and gaseous exchange.
   d. Blood cell production, energy storage and waste elimination.

2. **The production of blood cells and platelets takes place in:**
   a. Red bone marrow
   b. Yellow bone marrow
   c. The heart
   d. The liver.

3. **Mature cells that maintain the daily activities of bone tissue are called:**
   a. Osteoprogenitor cells
   b. Osteoblasts
   c. Osteoclasts
   d. Osteocytes.

4. **A hard bone tissue that is composed of a basic structural unit called an osteon, or Haversian system, is:**
   a. Compact bone tissue
   b. Spongy bone tissue
   c. Immature bone tissue
   d. Mature bone tissue.

5. **What is the name of the process through which most bones are formed?**
   a. Remodelling
   b. Desquamation
   c. Ossification
   d. Demineralisation.

6. **Which of the following statements is not true?**
   a. Bone is a dynamic, living tissue which is constantly changing, repairing and reshaping itself.
   b. New bone tissue constantly replaces old, worn-out or injured bone tissue.
   c. Bones are dead materials that support and protect the body and allow for movement.
   d. Ageing causes the bones to lose calcium and other minerals from its matrix.

7. **Which of the following are all different types of bones?**
   a. Long, flat, short and sesamoid.
   b. Long, short, broad and bursae.
   c. Bursae, irregular, flat and broad.
   d. Broad, short, sesamoid and long.

8. **An example of a long bone is the:**
   a. Patella
   b. Scapula
   c. Triquetrum
   d. Femur.

9. **What type of bone is a vertebra?**
   a. Flat
   b. Short
   c. Irregular
   d. Broad.

10. **Which of the following statements is true?**
   a. Flat bones act as areas of attachment for skeletal muscles and also provide protection.
   b. Long bones are hard and straight and function in protection.
   c. Bursae are complex bones that act as areas of attachment for skeletal muscles.
   d. Sesamoid bones are long and slightly curved to provide strength for locomotion.

11. **In an adult bone, where is yellow bone marrow found?**
   a. Epiphysis
   b. Diaphysis
   c. Periosteum
   d. Medullary.

12. **Which of the following bones are all categorised as part of the axial skeleton?**
   a. Hyoid, ribs and femur.
   b. Sternum, vertebrae and maxillae.
   c. Auditory ossicles, ulna and pisiform.
   d. Tibia, mandible and phalanges.

13. **What is the name of the two tiny bones that form the medial wall of the eye orbit?**
   a. Maxillae
   b. Lacrimals
   c. Temporals
   d. Parietals.

14. **What is the correct name for the lower jaw bone?**
   a. Sphenoid
   b. Ethmoid
   c. Maxilla
   d. Mandible.

15. **Which of the following are all bones of the hand?**
   a. Trapezoid, lunate, navicular, pisiform.
   b. Cuboid, trapezium, capitate, hamate.
   c. Calcaneus, cuneiform, pisiform, hamate.
   d. Capitate, hamate, scaphoid, lunate.

5 The Muscular System

Revision/Self-study notes

The essential function of all muscles is to shorten themselves, or contract. In doing so, they are able to perform a variety of functions in the body.

1. List the functions of the muscular system.

- _____

- _____

- _____

- _____

2. There are three types of muscle tissue in the body – skeletal, cardiac and smooth (visceral). Each type of muscle has a specific anatomy, location and function. Make up a mnemonic or draw an image for each muscle type so that you will always be able to remember the following information:

- **Skeletal muscle** – striated fibres; attached to bones, skin or other muscles; functions in locomotion, movements of lymph and venous blood, posture and heat production; under voluntary control.

Mnemonic: _____

- **Cardiac muscle** – striated fibres; forms most of the heart, pumps blood and helps regulate blood pressure; involuntary control.

Mnemonic: _____

- **Smooth (visceral) muscle** – non-striated fibres; forms walls of hollow internal structures; moves substances through tracts and regulates organ volume; involuntary control.

Mnemonic: _____

3. The figure below shows the different connective tissues associated with a skeletal muscle. Write a short phrase or a few words near to each label that will remind you of its function.

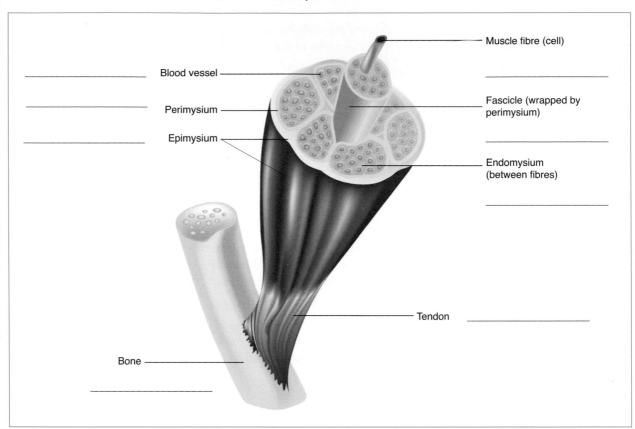

Muscle fibre (cell)

Blood vessel

Perimysium

Epimysium

Fascicle (wrapped by perimysium)

Endomysium (between fibres)

Tendon

Bone

4. When studying the muscular system it helps if you can learn some of the terminology associated with this system. In the table below give the definition for each term.

TERM	DEFINITION
Myofibre	
Sarcolemma	
Sarcoplasm	
Myofibril	
Myofilament	
Sarcomere	
Sarcoplasmic reticulum	

5. In your own words describe how a muscle contracts and relaxes. Make
   sure you include the following terms or details and cross each one off the
   list as you write about it.

   • sarcomere • thick filament • thin filament • myosin • myosin heads/cross-bridges • actin
   • myosin-binding site • elastic filaments • A-band • I-band • Z-disc • sliding-filament mechanism • ATP
   • nerve impulse • muscle-action potential • acetylcholine • calcium • power-stroke.

   _____

   _____

   _____

   _____

   _____

   _____

   _____

6. Physical activities include two types of contraction – isotonic and
   isometric. Complete the table below which highlights the differences
   between these two types of contraction.

DESCRIPTION	ISOTONIC	ISOMETRIC
Does muscle shorten?		
Is movement created?		
Function of contraction?		
What types of cells does this layer contain?		
Write a few words or a short phrase to remind you of this type of contraction.		

Study groups

Remember that when learning anatomy and physiology you are learning about yourself. So bring the subject to life by performing the different types of muscular contractions. One person in a group can perform the following types of contractions: isotonic (concentric and eccentric) and isometric. As the person performs each action the other members of the group need to say what type of contraction is being performed.

7. Muscles need energy in the form of ATP to contract and they gain this through either the phosphagen system or glycolysis. Write a short phrase or a few words to describe each of the following processes:

- The phosphagen system _____

- Glycolysis _____

- Anaerobic glycolysis _____

- Aerobic respiration _____

8. Three types of muscle fibres exist – slow oxidative, fast oxidative and fast glycolytic. Write a mnemonic to remind you of the characteristics of each:

- **Slow oxidative** – red, small, good oxygen supply, generate ATP aerobically and split it slowly, resistant to fatigue, plentiful in muscles of endurance or posture.

Mnemonic: _____

- **Fast oxidative** – red, medium sized, good oxygen supply, generate ATP aerobically and split it quickly, less resistant to fatigue, plentiful in muscles for walking and running.

Mnemonic: _____

- **Fast glycolytic** – white, large, poor oxygen supply, generate ATP anaerobically and split it very fast, fatigue easily, plentiful in muscles for fast movements.

Mnemonic: _____

9. **Define the following:**

- Origin ....

- Insertion..

- Prime mover (agonist)...

- Antagonist...

- Synergist..

- Fixator (stabiliser)..

10. Below and on the next page are diagrams of the anterior and posterior muscles of the body. Try to label as many as you can. As you label each muscle, try to memorise its basic action. To help you complete this task, refer to pages 160–185 of *Anatomy and Physiology for Beauty and Complementary Therapies*.

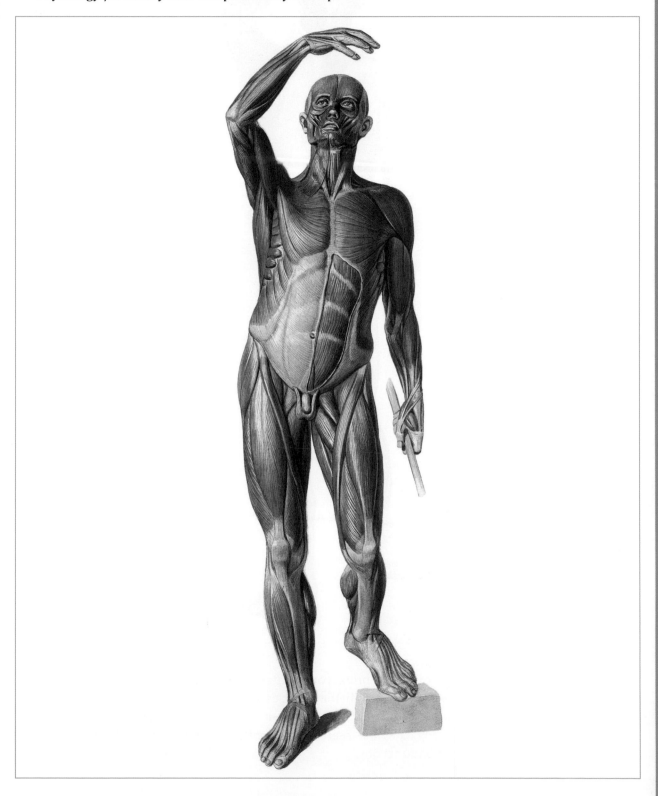

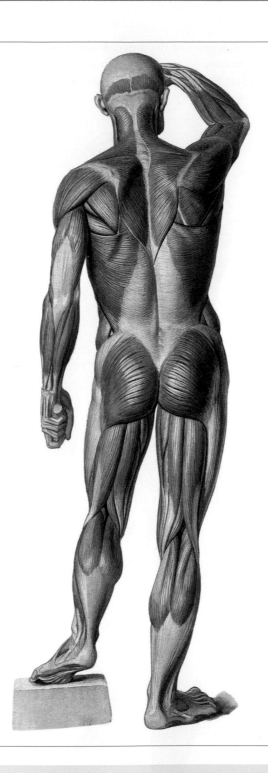

Studytip

Once again, bring your anatomy and physiology studies to life by working through the muscles on your own body. As you learn a particular muscle, find it on your own body. Try to feel it and find where it originates and inserts. Try to contract the muscle so that you can learn its basic actions.

Common pathologies of the muscular system

As a therapist you will need to have a basic understanding of a wide variety of muscle diseases and disorders. Write down a short phrase that will remind you of each of the following.

Diseases and disorders of muscles, bursae and tendons

- Fibromyalgia _____

- Fibrositis (muscular rheumatism) _____

- Ganglion cyst _____

- Muscular dystrophies _____

- Myasthenia gravis _____

- Poliomyelitis (polio) _____

- Rupture _____

- Spasm _____

- Strain _____

- Tetanus (lockjaw) _____

Sports injuries and repetitive strain injury (RSI)

- Carpal tunnel syndrome _____

- Frozen shoulder _____

- Golfer's elbow (medial epicondylitis) _____

- Housemaid's knee (pre-patellar bursitis) _____

- Shin splints _____

- Tendonitis _____

- Tennis elbow (lateral epicondylitis) _____

Exercises

1. Complete the crossword below.

Across:
3. The main energy-transferring molecule in the body (3).
4. The basic functional unit of a skeletal muscle (9).
7. Outermost layer of connective tissue that encircles the whole muscle (9).
10. What sugar is a major energy source for muscles? (7)
11. The connective tissue that surrounds each individual muscle fibre within a fascicle (10).
14. A function of the muscular system (10).
16. The generation of heat in the body (13).
17. A protein found in thin muscle filaments (5).
18. What compound, necessary for muscular contraction, can be stored in the muscles as myoglobin or stored in the blood as haemoglobin? (6)

Down:
1. Mineral necessary for muscle contraction to occur (7).
2. The term used to describe the partial contraction of a resting muscle (4).
5. If a muscle is overstimulated it can become weak and no longer respond to stimulus. This is called muscle _____ (7).
6. Another name for smooth muscle (8).
8. The plasma membrane of a muscle cell (10).
9. Another term for a muscle cell or muscle fibre (8).
12. The ability of muscles to return to their original shape after contracting or extending (10).
13. What type of muscle forms most of the heart? (7)
15. A lack of muscle tone (5).

2. Are the following statements true or false? Write T or F in the spaces provided.

a. Smooth muscle is non-striated, involuntary and pumps blood. ____

b. Cardiac muscle is striated and involuntary. ____

c. Skeletal muscle maintains posture and generates heat. ____

d. Skeletal muscle forms the walls of hollow internal structures such as the gastrointestinal tract. ____

e. Fascia, epimysium, perimysium and endomysium are all types of connective tissue. ____

f. Another term for 'muscle cell' is 'myofilament'. ____

g. Muscles contract through what is known as the sliding filament mechanism. ____

h. Sarcomeres contain filaments that move to overlap one another and cause a muscle to shorten. ____

i. Skeletal muscles are described as striated because they are made up of bands of thick and thin filaments that appear as light and dark striations. ____

j. Thick filaments contain molecules of the proteins actin and titin. ____

3. Fill in the missing words.

a. There are two types of muscular contraction, _____ and _____ . In an _____ contraction, muscles shorten and create movement while the tension in the muscle remains constant. In an _____ contraction, muscles contract but do not shorten and no movement is generated.

b. _____ contractions improve strength and joint mobility while _____ contractions stabilise joints and improve muscle tone.

c. The difference between aerobic and anaerobic processes is that aerobic processes require the presence of _____ .

4. There are three different types of skeletal muscle fibres – slow oxidative, fast oxidative and fast glycolytic. Complete the table below.

FEATURES	SLOW OXIDATIVE	FAST OXIDATIVE	FAST GLYCOLITIC
Colour	Red	Red to pink	c.
Diameter	a.	Medium	Largest
Activities	Maintains posture	b.	Fast movements such as ball throwing

5. The figure below shows the relationship of skeletal muscles to bones. Put the following labels in their correct place.

1. Belly 2. Attachment/Insertion 3. Fascia 4. Origin 5. Tendon

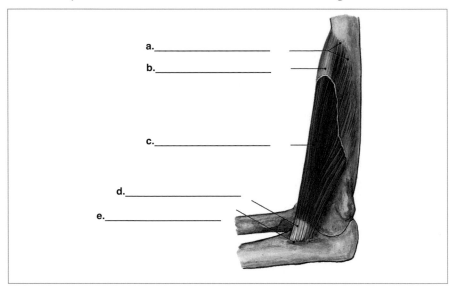

a._____

b._____

c._____

d._____

e._____

6. Listed below are descriptions of some of the muscles of the face, scalp and neck. Name each muscle described.

a. Found at the base of the skull and pulls the scalp backwards when raising the eyebrows or wrinkling the forehead.

_____

b. Circles the eye and is responsible for closing it.

_____

c. Circles the mouth and moves the lips during speech.

_____

d. Moves the cheeks when sucking or blowing.

_____

e. Originates in the mandible, inserts into the chin and is responsible for pulling up the chin or wrinkling it.

_____

f. Shaped like a strap and plays an important role in moving the head. It originates in the sternum and clavicle and inserts in the mastoid process.

_____

7. In the list below, circle only the muscles that are located on the anterior thorax and abdomen.

- Erector spinae
- Pectoralis minor
- Rhomboideus major
- Serratus anterior
- Trapezius
- Splenius capitis
- Rectus abdominis
- Infraspinatus
- Pectoralis major

8. Below are some of the muscles of the body. Connect the muscle to the area to which it belongs.

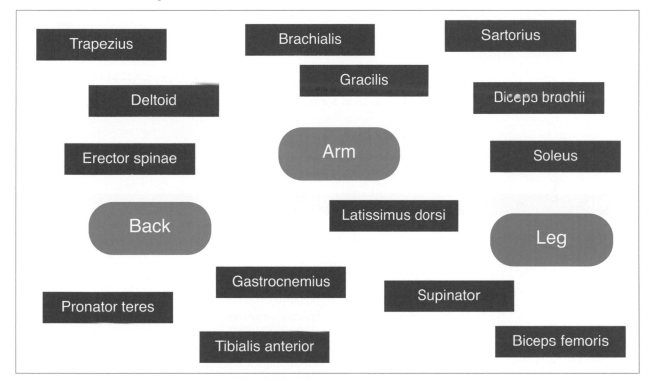

9. In the table below, the incorrect action for each muscle has been given. Match the correct action to each muscle by matching the number to the letter. Write your answers in the spaces provided.

TERM	DESCRIPTION
1 Soleus	a Flexes middle phalanges of each finger.
2 Extensor carpi ulnaris	b Dorsiflexes and inverts foot and extends big toe.
3 Tibialis anterior	c Plantar flexes and everts the foot.
4 Extensor hallicus longus	d Abducts the thumb.
5 Abductor pollicis brevis	e Plantar flexes the foot.
6 Peroneus longus	f Extends and adducts wrist.
7 Flexor digitorum superficialis	g Dorsiflexes the foot.

Answers: 1. ____ 2. ____ 3. ____ 4. ____ 5. ____ 6. ____ 7. ____

Common pathologies of the muscular system

10. Are the following statements true or false? Write T or F in the spaces provided.

a. An involuntary muscle contraction is called a spasm. ____

b. Poliomyelitis is a fungal infection. ____

c. Fibromyalgia is characterised by aching stiffness and pain in the soft tissue. ____

d. Myasthenia gravis and muscular dystrophy are both types of arthritis. ____

e. Carpal tunnel syndrome is a disorder of the foot. ____

f. The clinical term for tennis elbow is pre-patellar bursitis. ____

Vocabulary test

Complete the table below.

WORD	DEFINITION
a. _____	A flat, sheet-like tendon that attaches muscles to bone, skin or another muscle.
b. _____	The wasting away of muscles.
c. _____	Connective tissue that surrounds and protects organs, lines walls of the body, holds muscles together and separates muscles.
d. _____	The replacement of connective tissue by scar tissue.
e. _____	An increase in muscle tone.
f. _____	The point where a muscle attaches to the moving bone of a joint.
g. _____	The point where a muscle attaches to the stationary bone of a joint.
h. _____	A strong cord of dense connective tissue that attaches muscles to bones, to the skin or to other muscles.

Multiple choice questions

1. Which of the following is a function of the muscular system?
   a. Excretion
   b. Thermogenesis
   c. Ingestion
   d. Metabolism.

2. Which type of muscle tissue contracts to help return venous blood to the heart, move lymph through the lymphatic vessels, maintain posture and generate heat?
   a. Smooth
   b. Skeletal
   c. Cardiac
   d. Visceral.

3. What types of filaments do sarcomeres contain?
   a. Thin, thick and elastic.
   b. Short, long and wide.
   c. Black, white and red.
   d. Red, white and pink.

4. Which mineral is necessary for muscle contraction to occur?
   a. Carbon dioxide
   b. Zinc
   c. Oxygen
   d. Calcium.

5. Which type of muscle fibres are white, large, have a poor oxygen supply, produce ATP anaerobically and are plentiful in muscles used for fast movements such as throwing a ball?
   a. Slow oxidative
   b. Fast oxidative
   c. Slow glycolytic
   d. Fast glycolytic.

6. Which of the following statements is true?
   a. When muscles contract they shorten and usually move the moving bone towards the stationary bone.
   b. The prime mover muscle always opposes the antagonist.
   c. Fixators oppose the movements of both prime movers and antagonists.
   d. Another term for prime mover is antagonist.

7. Which of the following is not a muscle of facial expression?
   a. Occipitalis
   b. Nasalis
   c. Serratus anterior
   d. Procerus.

8. Which muscle originates in the temporal and frontal bones, inserts into the mandible and helps to move the mandible?
   a. Buccinator
   b. Risorius
   c. Sternocleidomastoid
   d. Temporalis.

9. What muscle forms the floor of the thoracic cavity and functions in inhalation?
   a. Internal intercostals
   b. Diaphragm
   c. Levator scapulae
   d. Transversus abdominis.

10. What is the main function of the rotator cuff muscles?
    a. To stabilise the abdominal muscles.
    b. To stabilise the spinal column.
    c. To hold the head of the femur in the cavity of the ilium.
    d. To hold the head of the humerus in the cavity of the scapula.

11. What structure does the flexor pollicis brevis muscle move?
    a. Neck
    b. Spinal column
    c. Thumb
    d. Big toe.

12. Which of the following is not a muscle located in the buttocks?
    a. Soleus
    b. Gluteus maximus
    c. Gluteus medius
    d. Gluteus minimus.

13. **What is the correct term used to describe the tearing of muscle fascia?**
    a. Sprain
    b. Strain
    c. Cramp
    d. Rupture.

14. **What is the name of the cellular process through which glucose is split into pyruvic acid and ATP?**
    a. Hypertonia
    b. Hypotonia
    c. Glycolysis
    d. Thermogenesis.

15. **What is the thenar eminence?**
    a. An area of soft tissue found on the ulnar side of the palm.
    b. An area of soft tissue found on the radial side of the palm.
    c. An area of soft tissue found beneath the little finger.
    d. An area of soft tissue found in the centre of the palm.

6 The Nervous System

Revision/Self-study notes

1. The nervous system has three main functions – sensory, integrative and motor. For each of these functions write a short sentence to demonstrate to yourself that you understand these functions.

• Sensory _____

• Integrative_____

• Motor _____

Organisation of the nervous system

Because the nervous system is such a complex system it can be divided into many smaller systems. It is very important to know what these different systems are and how they function together as one coordinated whole. In the space on the next page draw your own spider diagram/mind-map that shows how these divisions connect. The divisions are:

• **Central nervous system (CNS)** – This system is composed of the brain and spinal cord and functions in processing and integrating information.
• **Peripheral nervous system (PNS)** – This system is composed of the cranial and spinal nerves and it functions in transporting impulses between the CNS and the rest of the body.
• **Somatic nervous system (voluntary nervous system)** – This system is a subdivision of the PNS. It allows us to control our skeletal muscles.
• **Autonomic nervous system** – Also called the involuntary nervous system, this system is a subdivision of the PNS. It controls all automatic processes of smooth muscle, cardiac muscle and glands.
• **Sympathetic nervous system** – This division falls under the autonomic nervous system. It stimulates activity.
• **Parasympathetic nervous system** – This division falls under the autonomic nervous system and it works in opposition to the sympathetic nervous system by inhibiting activity.

**Study**tip

Make your spider diagram or mind map as personal as possible and have fun doing it! You will always find it easier to remember colours, symbols, diagrams or even rhymes or jokes that you have chosen. As you create the spider diagram or mind map you will begin to understand the complexities of the nervous system.

Nervous tissue

1. The nervous system contains two types of cells – neuroglia and neurones. In the spaces provided write the basic function of each of these cells:

• Neuroglia _____

_____

• Neurones _____

_____

2. Neurones can be classified according to the type of information they carry and the direction in which they carry that information. Complete the table on the next page.

DESCRIPTION	AFFERENT NEURONE	EFFERENT NEURONE
Alternative name (i.e. sensory or motor)		
Type of information carried (i.e. sensory or motor nerve impulse)		
Direction carried (i.e. from and to)		
Write a few words or a short phrase to remind you of this type of neurone. For example: Afferent = Arrive Efferent = Exit		

There is a third type of neurone which has not been included in this table. This is an association or inter-neurone. What is the function of this type of neurone?

_____

3. The diagram below shows the structure of a typical motor neurone. Label the diagram with the following:

• Cell body • Dendrites • Axon • Axon terminal
• Synaptic vesicle • Myelin sheath • Nodes of Ranvier

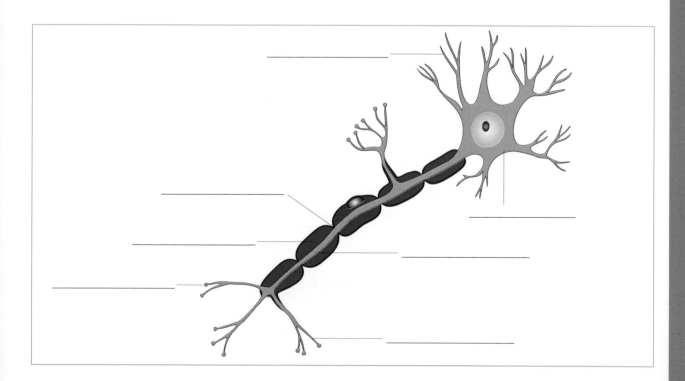

4. Learning how a nerve impulse is transmitted is quite a difficult concept to grasp. To double-check that you really do understand how nerves are transmitted, put yourself in the examiner's seat and write four questions asking about nerve transmission. Then put yourself back into your own seat and answer the questions.

Studytip

You may find it easier and more interesting if you do this exercise as part of a group.

Question 1:_____

Answer:_____

_____

_____

Question 2:_____

Answer:_____

_____

_____

Question 3:_____

Answer:_____

_____

_____

Question 4:_____

Answer:_____

_____

_____

5. A nerve consists of a bundle of nerve fibres surrounded by connective tissue. This connective tissue includes endoneurium, perineurium and epineurium. Describe each of these.

• Endoneurium _____

• Perineurium _____

• Epineurium _____

Studytip

Compare the connective tissue coverings of nerves to those of muscles.

The brain

1. The figure below shows the human brain. Colour in each region of the brain and next to the colour key write 3–5 words that will remind you of the functions of that region.

Try to choose colours that are significant to you and as you are colouring in each region go through in your mind the different functions of that region.

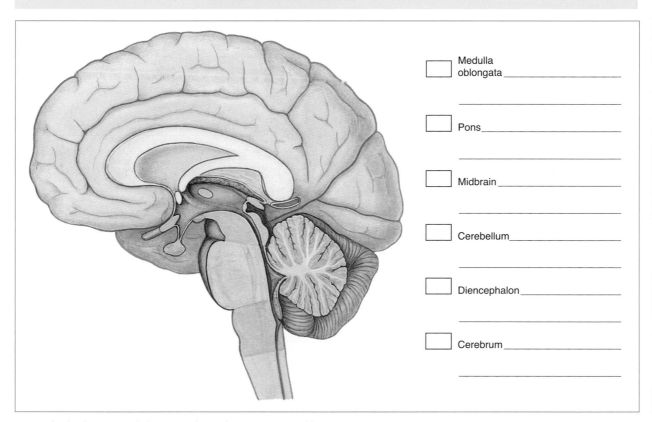

Medulla oblongata _____

_____

Pons_____

_____

Midbrain _____

_____

Cerebellum_____

_____

Diencephalon_____

_____

Cerebrum_____

_____

2. Both the brain and the spinal cord are protected by connective tissue coverings called meninges. In the space provided write a mnemonic to remind you of the location of each meninge:

- Dura mater – outer covering
- Arachnoid – middle covering
- Pia mater – inner covering

Mnemonic: _____

3. The brain and spinal cord are also bathed in a unique fluid called cerebrospinal fluid. List its functions:

- _____

- _____

- _____

Cranial nerves

1. List the cranial nerves and write a short phrase or a few words to remind you of their functions.

I _____

II _____

III _____

IV _____

V _____

VI _____

VII _____

VIII _____

IX _____

X _____

XI _____

XII _____

Spinal cord

1. Describe the two main functions of the spinal cord.

• _____

• _____

2. List the differences between the grey matter and white matter of the nervous system.

	GREY MATTER	WHITE MATTER
Structure		
Function		

3. There are 31 pairs of nerves that originate in the spinal cord and they all
   form part of the PNS because they connect the CNS to muscles or glands.
   Branches of some of these nerves form networks called plexuses that
   innervate specific structures of the body. Complete the table below.

PLEXUS	AREA INNERVATED
Cervical	
Brachial	
Lumbar	
Sacral	

Special sense organs

1. Our eyes are one of the brain's most vital contacts with the outside world.
   However, many students find it difficult to study these fascinating
   structures because of the terminology involved. So to help you learn about
   the eye, make cue cards of the following terms and their definitions:

Lacrimal glands	Conjunctiva	Fibrous tunic
Cornea	Sclera	Vascular tunic
Iris	Ciliary body	Choroid
Photoreceptors	Vitreous body	Lens
Cones	Rods	Aqueous humour
Nervous tunic	Pigmented portion of nervous tunic	Neural portion of nervous tunic
Posterior cavity/ vitreous chamber	Anterior cavity	

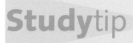 **Study**tip — Write the definitions in your own words. It is always easier to remember your own words than someone else's.

2. Our ears enable us to hear, locate sounds and maintain our balance. They can be divided into three regions – the external or outer ear that we see, the middle ear and the internal or inner ear. In order to learn these different regions of the ear draw the 'journey' of a sound wave in the space provided below. The first few steps have been done for you.

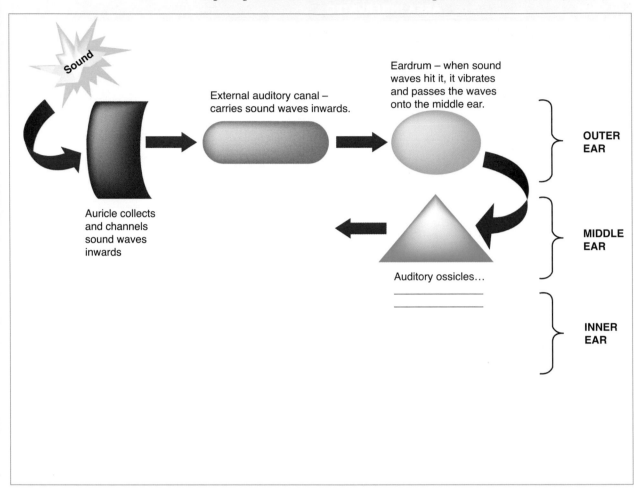

3. The senses of taste and smell are closely connected. In the spaces below summarise the physiology behind these two senses.

- Sense of taste _____

  _____

  _____

  _____

- Sense of smell _____

  _____

  _____

  _____

Common pathologies of the nervous system

Write down a short phrase that will remind you of each of the following diseases and disorders.

Nervous system diseases and disorders

• Alzheimer's disease _____

• Anosmia _____

• Brain tumour _____

• Cerebral palsy _____

• Cerebrospinal accident (CVA, Stroke) _____

_____

• Depression _____

• Epilepsy _____

• Fainting (syncope)_____

• Insomnia _____

• Meningitis _____

• Motor neurone disease (MND)_____

• Myalgic encephalomyelitis (ME, Chronic fatigue) _____

_____

• Pain and referred pain (synalgia) _____

• Parkinson's disease (PD) _____

• Spina bifida _____

• Stress_____

• Cluster headaches _____

• Migraines_____

• Spinal cord injuries_____

• Tension headaches _____

Nerve disorders

• Bell's palsy _____

• Multiple sclerosis _____

• Neuritis _____

• Trigeminal neuralgia_____

Head injuries
- Concussion _____

- Contusion _____

- Laceration _____

Eye disorders
- Cataract _____

- Conjunctivitis (pink eye) _____

- Corneal ulcer _____

- Glaucoma _____

- Stye (hordeolum) _____

Ear disorders
- Deafness _____

- Earache _____

- Glue ear (secretory or serous otitis media) _____

- Otitis media _____

- Tinnitus _____

- Vertigo _____

Exercises

1. The nervous system can be divided into the Central Nervous System
   (CNS) and Peripheral Nervous System (PNS). Complete the spider
   diagram below, using arrows to connect the structures and functions to
   where they belong, either the CNS or PNS.

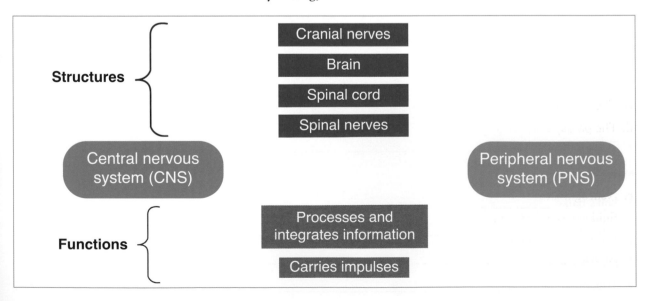

2. **The Peripheral Nervous System can also be divided into subdivisions. Complete the diagram below by filling in the missing words on the lines provided.**

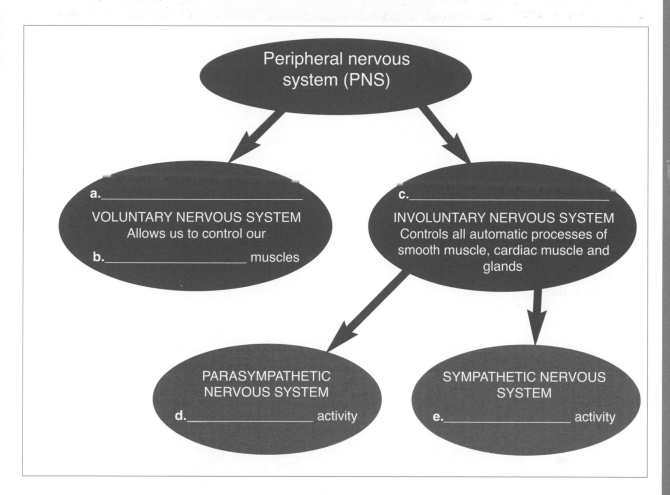

3. **Are the following statements true or false? Write T or F in the spaces provided.**

a. Afferent neurones conduct impulses from sensory receptors to the CNS. _____

b. Motor neurones are also called afferent neurones. _____

c. The sympathetic nervous system decreases the heart beat. _____

d. The parasympathetic nervous system opposes the actions of the sympathetic nervous system by inhibiting activity and conserving energy. _____

e. Neuroglia are the cells responsible for maintaining the homeostasis of the fluid surrounding neurones. _____

f. Efferent neurones transmit information from the skin, sense organs, muscles, joints and viscera to the CNS. _____

4. All neurones have three parts – a cell body, dendrites and an axon terminal. Complete this table on the structure of a typical motor neurone.

STRUCTURE	FUNCTION
a.	The metabolic centre of a neurone.
Dendrite	b.
Axon	c.
d.	Chemical that influences other neurones, muscle fibres or gland cells.
e.	Neurotransmitter-containing sac found at the end of axons.
Myelin sheath	f.
g.	Cells that form the myelin sheath.
Nodes of Ranvier (neurofibril nodes)	h.

5. The figure below is an overview of the brain. Complete the labelling.

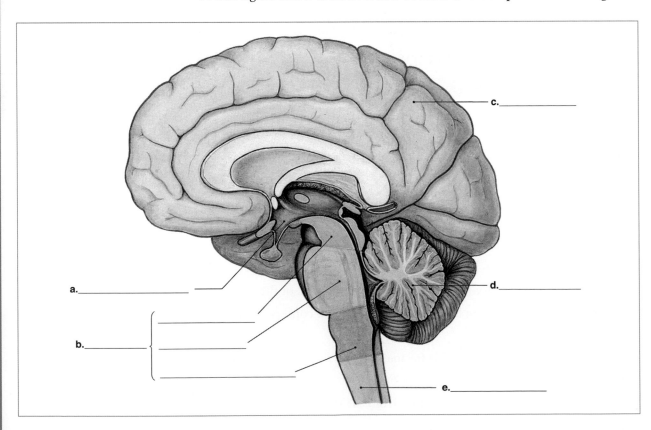

a._____

b._____

c._____

d._____

e._____

6. Use the clues given to unscramble the letters below.

      a. RDUAAETMR   **b.** EICNVTESLR   **c.** EMINESNG   **d.** RAHDAOCNI

Answers: _____   _____   _____   _____

Clues:
- These three layers cover the brain and help protect it from the external environment.
- The outer connective tissue covering of the brain.
- The middle connective tissue covering of the brain.
- Cavities in the brain through which CSF can flow.

7. Use the clues below to complete this wordsearch.

L	I	M	B	I	C	S	Y	S	T	E	M	D	F	H	C
D	G	J	T	X	A	D	W	I	G	E	I	G	A	F	E
S	D	G	W	C	E	R	E	B	R	U	M	C	N	A	R
D	D	L	E	E	O	D	T	G	I	O	D	E	S	E	E
E	O	Q	B	R	A	I	N	S	T	E	M	O	W	D	B
P	Y	F	O	E	H	E	D	H	W	P	K	H	T	H	R
H	R	J	T	B	D	N	O	D	E	D	G	V	E	O	A
Y	H	F	F	E	J	C	T	U	L	E	S	A	U	I	L
F	V	C	A	L	D	E	G	I	V	U	N	F	T	G	C
K	M	O	W	L	X	P	F	R	E	S	C	J	S	J	O
V	V	A	G	U	S	H	A	P	Q	U	W	D	T	N	R
E	B	T	X	M	X	A	B	D	U	C	E	N	S	E	T
J	Y	V	D	F	E	L	S	O	T	T	O	G	R	C	E
U	I	D	X	A	T	O	L	F	A	C	T	O	R	Y	X
E	R	J	D	D	R	N	G	S	F	G	T	A	G	B	I

Clues:
- Connects the spinal cord to the diencephalon.
- Area of the brain responsible for movement, posture and balance. Found at the back of the head.
- Houses the pituitary and pineal endocrine glands, also contains the epithalamus, thalamus and hypothalamus. Sometimes called the interbrain.
- Area of the brain that gives us the ability to read, write, speak, remember, create and imagine. It is often referred to as the 'seat of intelligence'.
- The 'emotional brain'. Area of the brain associated with pain and pleasure as well as memory.
- The cerebrum's outermost layer. It is made up of grey matter.
- How many pairs of cranial nerves are there?
- Cranial nerve I. This nerve functions in the sense of smell.
- Cranial nerve VI. This nerve helps control the movement of the eyeball.
- Cranial nerve X. This nerve helps control the secretion of digestive fluids and the contraction and relaxation of the smooth muscle of the organs of the thoracic and abdominal cavities.

8. As part of the central nervous system, the spinal cord is integral to maintaining the homeostasis of the body. The columns in the table below are jumbled. Match the numbers and letters and write your answers in the spaces provided.

STRUCTURE	FUNCTION
1 White matter	**a** A network of nerves.
2 Cauda equina	**b** Neurological tissue that contains myelinated neurones and functions in transporting impulses.
3 Grey matter	**c** The thin, inner connective tissue covering of the spinal cord.
4 Plexus	**d** A bundle-like structure of nerves that extends from the end of the spinal cord.
5 Pia mater	**e** Neurological tissue that contains unmyelinated neurones and functions in receiving and integrating information.

Answers: 1. _____ 2. _____ 3. _____ 4. _____ 5. _____

9. The figure below shows the structure of the eyeball. Label the following areas:

1. Iris   2. Cornea   3. Lens   4. Anterior cavity

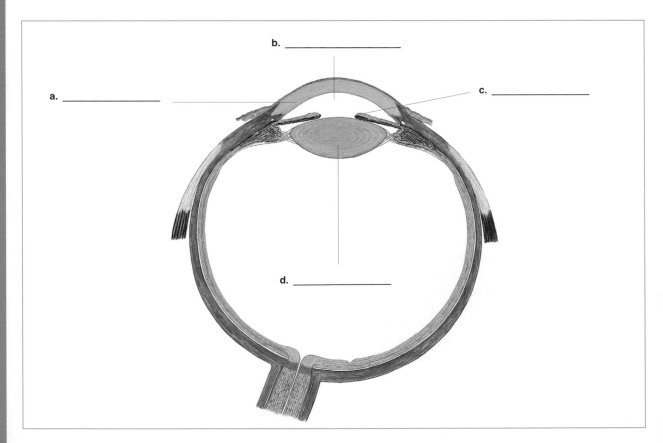

10. Fill in the missing words.

a. The _____ ear collects and channels sound waves inwards. The part of the ear that we see is called the _____ . The eardrum is also called the _____ membrane and when sound waves hit it, it _____, passing the waves on to the _____ ear.

b. The _____ ear is a small air-filled cavity that is partitioned from the outer ear by the _____ and the inner ear by the _____ window and the _____ window. The middle ear contains three tiny bones, the _____ , _____ and _____ . Together these bones are called the _____ _____ . The middle ear is connected to the throat by the _____ _____ .

c. The _____ ear is sometimes also called the labyrinth and it consists of a bony labyrinth filled with a fluid called _____ enclosing a membranous labyrinth filled with a fluid called _____ . The labyrinth is divided into three areas – the _____ , the three semicircular _____ and the organ of _____ .

11. Are the following statements true or false? Write T or F in the spaces provided.

a. The receptors of taste are called gustatory receptors and they are found in taste buds on the tongue, back of the roof of the mouth and in the pharynx and larynx. _____

b. Gustatory receptor cells are also called Bowman's glands. _____

c. Gustatory receptor cells are chemoreceptors. _____

d. The receptors of smell are called olfactory receptors. _____

e. The senses of taste and smell are linked and their impulses travel to the cerebellum in the brain. _____

Common pathologies of the nervous system

12. Give the names of the following pathologies of the nervous system.

1. Inflammation of the meninges that cover the brain and spinal cord.

_____

2. Disorders of the brain characterised by seizures which are short, recurrent, periodic attacks of motor, sensory or psychological malfunction.

_____

3. The loss of the sense of smell.

_____

4. The progressive degeneration of brain tissue. Symptoms include memory loss, forgetfulness, disorientation and confusion.

_____

5. A birth defect in which the spinal vertebrae do not form normally.

_____

6. Weakness or paralysis of the muscles on one side of the face.

_____

7. The inflammation of a nerve or a group of nerves.

_____

8. Inflammation of the sciatic nerve.

_____

9. The temporary loss of consciousness following a head injury.

_____

10. Loss of vision due to abnormally high pressure in the eye. A result of the eye's inability to drain aqueous humour as quickly as it produces it.

_____

Vocabulary test

Complete the table below.

WORD	DEFINITION
a. _____	A bundle or knot of nerve cell bodies.
b._____	The three connective tissue membranes that enclose the brain and spinal cord.
c. _____	The study of the nervous system.
d._____	A specialised nerve receptor located in muscles, joints and tendons that provides sensory information regarding body position and movements.

Multiple choice questions

1. The part of the nervous system that reacts to changes by stimulating activity is called the:
   a. Somatic nervous system
   b. Autonomic nervous system
   c. Sympathetic nervous system
   d. Parasympathetic nervous system.

2. How many different types of cells does the nervous system contain?
   a. One
   b. Two
   c. Three
   d. Four.

3. What type of neurone transmits impulses from the skin to the Central Nervous System?
   a. Afferent
   b. Efferent
   c. Association
   d. Inter-neurones.

4. What type of neurone transmits impulses from the Central Nervous System to muscles?
   a. Afferent
   b. Efferent
   c. Association
   d. Inter-neurones.

5. What is another name for afferent neurones?
   a. Sensory neurones
   b. Motor neurones
   c. Mixed neurones
   d. Inter-neurones.

6. Where on a motor neurone are the synaptic vesicles found?
   a. In the myelin sheath.
   b. At the end of the dendrites.
   c. At the end of the cell body.
   d. At the end of the axons.

7. How are nerve impulses transmitted?
   a. Electronically
   b. Chemically
   c. Electrochemically
   d. None of the above.

8. What are the two ions necessary for the transmission of a nerve impulse?
   a. Sodium and potassium
   b. Calcium and magnesium
   c. Potassium and calcium
   d. Magnesium and sodium.

9. **Which of the following statements is true?**
   a. The three meninges are the outer pia mater, the middle arachnoid and the inner dura mater.
   b. The brain stem consists of the medulla oblongata, cerebellum and diencephalon.
   c. The cerebrum consists of the cerebral cortex which is the white matter of the brain.
   d. The brain is protected by the bones of the cranium, the meninges and cerebrospinal fluid.

10. **Which of the following is not a region of the brain?**
   a. Brain stem
   b. Diencephalon
   c. Pineal
   d. Cerebrum.

11. **Which of the following are all cranial nerves?**
   a. Optic, trigeminal, facial, accessory.
   b. Olfactory, brachial, abducens, trochlear.
   c. Hypoglossal, cervical, lumbar, vagus.
   d. Thoracic, oculomotor, vestibulocochlear, glossopharyngeal.

12. **Approximately where in the spinal column does the spinal cord end?**
   a. C3
   b. T2
   c. L2
   d. S3.

13. **How many pairs of spinal nerves are there?**
   a. 11
   b. 21
   c. 31
   d. 41.

14. **Which of the following areas of the body is not served by the lumbar plexus?**
   a. Abdominal wall
   b. Buttocks
   c. External genitals
   d. Lower limb.

15. **Which of the following statements is true?**
   a. Light waves are bent as they pass through the structures of the eyeball. They then hit the retina where they are converted into nerve impulses.
   b. Sound waves are transmitted as vibrations through the outer ear until they reach the middle ear where they are converted into nerve impulses.
   c. Tastes are chemicals that come into contact with hair-like projections on olfactory receptors where they are converted into nerve impulses.
   d. Smells are chemicals that dissolve in mucous in the nose. They then come into contact with taste buds and are converted into nerve impulses.

7 The Endocrine System

Revision/Self-study notes

Functions of the endocrine system

1. There are two types of glands in the body – endocrine and exocrine.
   Complete the table below to highlight the differences between these glands.

	ENDOCRINE GLAND	EXOCRINE GLAND
Where are substances secreted?		
What happens to the secretions?		
Examples of secretions:		
Write a few words or a short phrase to remind you of each type of gland		

2. Hormone secretion is controlled by signals from the nervous system
   (neural stimulation), chemical changes in the blood and other hormones.
   Describe each of these in your own words.

- Neural stimulation _____

  _____

- Chemical changes _____

  _____

- Hormonal stimulation _____

  _____

The endocrine glands and their hormones
The endocrine system is another of those systems in which you will find a
great deal of information that you simply have to remember. Once again,
the only way to really do this is to sit down and memorise it all. To help
you do this, firstly complete the tables on the next page and then make
cue cards of the information contained in those tables.

Studytip

Write the actions in your own words or sketch pictures if you find it
easier to remember images rather than words.

GLAND	HORMONE	ACTION
Anterior pituitary	Human growth hormone	
	Thyroid-stimulating hormone	
	Follicle-stimulating hormone	
	Luteinizing hormone	
	Prolactin	
	Adrenocorticotropic hormone	
	Melanocyte-stimulating hormone	
Posterior pituitary	Oxytocin	
	Antidiuretic hormone	
Pineal	Melatonin	
Thyroid	Thyroid hormone	
	Calcitonin	
Parathyroids	Parathormone	
Thymus	Thymosin	
Pancreatic islets	Glucagon	
	Insulin	
	Somatostatin	
Adrenal cortex	Mineralcorticoids	
	Glucocorticoids	
	Sex hormones	
Adrenal medulla	Adrenaline	
	Noradrenaline	
Ovaries	Oestrogens	
	Progesterone	
Testes	Testosterone	

Common pathologies of the endocrine system

1. Write down a short phrase that will remind you of each of the following diseases and disorders:

Pathologies of the adrenal glands

- Addison's disease _____

- Cushing's syndrome _____

Pathologies of the pancreatic islets

- Diabetes mellitus _____

Pathologies of the parathyroid glands

- Hypocalcemia (calcium deficiency) _____

Pathologies of the pineal gland

- Seasonal affective disorder (SAD) _____

Pathologies of the pituitary gland

- Acromegaly _____

- Diabetes insipidus _____

- Gigantism _____

Pathologies of the thyroid gland

- Goitre _____

- Graves' disease _____

- Hashimoto's disease _____

- Myxoedema (hypothyroidism) _____

- Thyrotoxicosis (hyperthyroidism) _____

Study group

To help you revise your knowledge of the endocrine system, use the cue cards you have just made and play the game 30 Seconds. See Chapter 1, page 10 to learn how to play.

Exercises

1. Complete the crossword below.

Across:

1. This endocrine gland secretes a hormone which is thought to contribute to setting the timing of the body's biological clock (6).
3. This gland releases a hormone which helps control blood calcium levels (11).
6. An endocrine gland that is also considered part of the nervous system (12).
7. A gland that secretes substances into ducts which then carry these substances into body cavities or to the outer surface of the body (8).
10. The male sex glands (6).
11. A chemical messenger that is transported by the blood and that regulates cellular activity (7).
13. These two glands are found above the kidneys (7).
14. This butterfly shaped gland is found wrapped around the trachea, just below the larynx (7).

Down:

1. The endocrine gland that secretes human growth hormone, luteinizing hormone and adrenocorticotropic hormone amongst other hormones (9).
2. This structure functions as part of both the digestive and endocrine systems and its endocrine function is to help regulate blood glucose levels (8).
4. A female sex gland (5).
5. A hormone-secreting gland (9).
8. A body function controlled and coordinated by hormones (12).
9. Which region of the adrenal glands releases hormones that function in the fight-or-flight response of the body and help the body cope with stress (7)?
12. A body system that works closely with the endocrine system to control all the processes that take place in the body (7).
14. This gland is located in the thorax, behind the sternum and between the lungs (6).

2. Are the following statements true or false? Write T or F in the spaces.

a. Hormones are chemical messengers that are secreted by exocrine glands. _____

b. Hormones coordinate body functions such as growth, development, reproduction, metabolism and homeostasis. _____

c. Hormones such as insulin and oxytocin are steroid hormones. _____

d. Hormone secretion is controlled by signals from the nervous system, chemical changes in the blood and other hormones. _____

e. One of the key differences between the nervous and endocrine systems is that the nervous system takes longer to act than the endocrine system does. _____

f. The time at which a child becomes sexually mature is called the menopause. _____

3. The columns below show the major endocrine glands (orange boxes) and the hormones they secrete (blue boxes). However, some of these hormones are in the wrong columns. Cross out the hormones that are in the incorrect columns and write their names in the correct columns.

PITUITARY	PINEAL	THYROID	PARATHYROIDS	THYMUS
Human growth hormone	Prolactin	Thyroid hormone	Parathormone	Adrenocortico-tropic hormone
Glucagon	Adrenaline	Thymosin	Melatonin	Sex hormones
Calcitonin		Thyroid stimulating hormone		

PANCREAS	ADRENAL CORTEX	ADRENAL MEDULLA	OVARIES	TESTES
Antidiuretic hormone	Mineral-corticoids	Follicle-stimulating hormone	Oestrogens	Luteinising hormone
Somatostatin	Noradrenaline	Testosterone	Oxytocin	Progesterone
	Glucocorticoids	Insulin	Melanocyte-stimulating hormone	

Common pathologies of the endocrine system

4. Fill in the missing words.

1. Addison's disease is caused by _____ of glucocorticoids and aldosterone.

2. _____ _____ is a disorder in which there is an elevation of glucose in the blood. Symptoms include increased thirst and urination.

3. Hypocalcemia is a deficiency of the mineral _____ .

4. The letters S.A.D. stand for _____ _____

   _____ which is characterised by depression, a lack of

   interest in one's usual activities, oversleeping and overeating.

5. Hypersecretion of human growth hormone can cause _____

   in adults or _____ in children.

6. Diabetes insipidus is caused by a lack of _____ hormone.

7. An enlarged thyroid gland is called a _____ .

8. Myxoedema is also called _____ and is caused

   by the hyposecretion of thyroid hormones in adults.

9. _____ disease is a hereditary autoimmune disorder in

   which the thyroid gland is continually stimulated to produce hormones.

10. Hashimoto's disease is a disorder of the _____ gland.

Vocabulary test

Complete the table below.

WORD	DEFINITION
Apoptosis	a. _____
Endocrinology	b. _____
Gonads	c. _____
Hypersecretion	d. _____
Hyposecretion	e. _____

Multiple choice questions

1. **Which of the following is not a characteristic of a hormone?**
   a. Chemical messenger.
   b. Secreted by exocrine glands.
   c. Transported in the blood.
   d. Regulates cellular activity.
2. **Steroid hormones include:**
   a. Sex hormones
   b. Insulin
   c. Oxytocin
   d. Amino-acid based hormones.
3. **Which of the following statements is correct?**
   a. Hormones are transported along nerve axons.
   b. Hormones generally act quicker than nerve impulses.
   c. Hormones only affect muscle cells.
   d. Hormones help coordinate metabolic activities, growth, development and reproduction.
4. **Which gland does the hypothalamus have its greatest influence on?**
   a. Pituitary
   b. Pineal
   c. Thyroid
   d. Thymus.

5. **Where in the body is the pituitary gland located?**
   a. In the head, at the base of the occipital lobe.
   b. In the head, behind the nose and between the eyes.
   c. In the throat, below the larynx.
   d. In the chest, between the lungs.

6. **Which of the following are all hormones secreted by the pituitary gland?**
   a. Follicle-stimulating hormone, oestrogen, melatonin.
   b. Adrenaline, antidiuretic hormone, parathormone.
   c. Human growth hormone, oxytocin, prolactin.
   d. Calcitonin, luteinizing hormone, melanocyte-stimulating hormone.

7. **What is the main action of oxytocin?**
   a. Stimulates the development of oocytes.
   b. Stimulates ovulation.
   c. Stimulates contraction of the uterus during labour.
   d. Has an antidiuretic effect.

8. **Which hormone increases blood pressure, dilates airways to the lungs, decreases the rate of digestion and increases blood glucose levels?**
   a. Insulin
   b. Progesterone
   c. Parathormone
   d. Adrenaline.

9. **What hormone does the pineal gland secrete?**
   a. Thymosin
   b. Melatonin
   c. Melanocyte-stimulating hormone
   d. Calcitonin.

10. **Which hormone is responsible for the development of masculine secondary sex characteristics?**
   a. Testosterone
   b. Oestrogen
   c. Progesterone
   d. Somatostatin.

11. **What is menopause characterised by?**
   a. The appearance of secondary sexual characteristics.
   b. The start of menstruation.
   c. The cessation of ovulation and menstruation.
   d. The development of a foetus.

12. **The hormone calcitonin is secreted by which endocrine gland?**
   a. Thyroid
   b. Parathyroid
   c. Thymus
   d. Pancreas.

13. **Adrenocorticotropic hormone is secreted by which endocrine gland?**
   a. Adrenals
   b. Pineal
   c. Pituitary
   d. Pancreas.

14. **Which of the following best describes glucocorticoids?**
   a. A group of hormones which regulate the mineral content of the blood.
   b. A group of hormones which regulate metabolism and help the body resist long-term stressors.
   c. A group of hormones which function in the body's fight-or-flight response.
   d. A group of hormones which slow absorption of nutrients from the gastrointestinal tract.

15. **Which of the following are all symptoms of Cushing's syndrome?**
   a. Increased thirst and urination, weight loss, fatigue and blurred vision.
   b. Confusion, memory loss, depression, muscle aches and spasms.
   c. The abnormal lengthening of the long bones of the arms and legs.
   d. Fatigue and excessive fat deposits on the face, torso and back.

8 The Respiratory System

Revision/Self-study notes

Structure and functions of the respiratory system

1. The primary function of the respiratory system is respiration, which is the exchange of gases. Complete the table below which highlights the differences between the various types of respiration.

DESCRIPTION	PULMONARY VENTILATION	EXTERNAL RESPIRATION	INTERNAL RESPIRATION	CELLULAR / RESPIRATION OXIDATION
	This is another term for breathing which includes both inspiration and expiration.			This is a metabolic reaction in which oxygen is used and carbon dioxide is formed as a by-product.
Where does this occur?	Lungs	Between lungs and blood	Between blood and tissue cells	In cells
Is oxygen gained or lost?	On inspiration oxygen is gained and on expiration oxygen is lost.			
Is carbon dioxide gained or lost?				
Write a few words or a short phrase to remind you of this process.				

2. The respiratory system takes in oxygen from the air we breathe and this air travels through a series of interconnecting passageways from the nose to the alveoli where gaseous exchange between the lungs and blood finally takes place.

   In order to learn the structures of the respiratory system draw the journey of air in the space provided on the next page. As you label each structure write down any important information that you will need to remember (for example, structure and function). The first step in this journey has been done for you.

Nose – Air breathed in through external nares (nostrils) is filtered, warmed and moistened. Nose also receives olfactory stimuli and acts as a resonating chamber for sound.

air

3. The lungs are an unusual organ in that they have two blood supplies:

- Pulmonary arteries bring deoxygenated blood to the lungs. Here the blood is oxygenated by the lungs and this oxygenated blood is then transported to the heart by the pulmonary veins.
- Bronchial arteries bring oxygenated blood to the lung tissue. Most of this blood is returned to the heart by the pulmonary veins, however, some of it drains into bronchial veins which then transport the blood to the heart via the superior vena cava.

In the space provided below, draw a spider diagram or mind map of the above information.

Physiology of respiration

1. Pulmonary ventilation, or breathing, is dependent on the existence of a pressure gradient between the pressure inside the lungs and outside them. To help you remember the physiology of respiration, make up mnemonics using the information given below. Before you do this, however, rewrite Boyle's Law in your own words to ensure you understand it.

- Boyle's Law = The pressure of a gas in a closed container is inversely proportional to the volume of the container.

    Rewrite this law in your own words: _____

    _____

    _____

    _____

- **Inspiration:** diaphragm and external intercostal muscles contract → this causes the thoracic cavity to increase in size → this causes the pressure inside the thoracic cavity to drop = this creates a vacuum which sucks air inwards.

Mnemonic _____

- **Normal expiration (a passive process):** muscles recoil to their natural size → this causes the thoracic cavity to decrease in size → therefore the atmospheric pressure is less than the pressure within the lungs = thus air moves out of the lungs.

    Mnemonic _____

- **Active expiration:** abdominal and internal intercostal muscles contract → this forces the diaphragm upwards → this reduces the size of the thoracic cavity → this reduces the volume of the lungs → this increases the pressure within the lungs = therefore air is forced out of the lungs.

Mnemonic _____

2. External respiration (pulmonary respiration) is the exchange of oxygen and carbon dioxide between the alveoli and blood. In your own words, summarise this process.

_____

_____

_____

_____

_____

Common pathologies of the respiratory system

Write down a short phrase that will remind you of each of the following diseases and disorders.

- Chronic obstructive pulmonary disease (COPD) _____

- Asthma_____

- Bronchitis _____

- Emphysema _____

Infectious and environmentally related diseases and disorders

- Asbestosis _____

- Cor pulmonale _____

- Cystic fibrosis _____

- Hayfever _____

- Hyperventilation_____

- Influenza _____

- Laryngitis_____

- Lung cancer_____

- Methicillin resistant staphylococcus aureus (MRSA) _____

- Pharyngitis_____

- Pleurisy _____

- Pneumonia_____

- Pneumothorax _____

- Pulmonary embolism _____

- Pulmonary fibrosis _____

- Rhinitis _____

- Sarcoidosis_____

- Severe acute respiratory syndrome (SARS) _____

- Sinusitis _____

- Tuberculosis (TB) _____

- Whooping cough (pertussis) _____

Exercises

1. The table below gives some of the terminology particular to the respiratory system. However, the terms and their definitions are muddled up. Match each term to its correct definition by matching the number to the letter. Write your answers in the spaces provided.

TERM	DEFINITION
1 Pulmonary ventilation	**a** The exchange of gases between the lungs and the blood.
2 External respiration	**b** Area formed by the alveoli where the exchange of gases occurs.
3 Internal respiration	**c** A series of passageways that allows air to reach the lungs – includes the nose, pharynx, larynx, trachea and bronchi.
4 Cellular respiration	**d** Another term for breathing – the act in which air is breathed into and out of the lungs.
5 Olfaction	**e** The exchange of gases between blood and tissue cells.
6 Conducting zone	**f** The metabolic reaction that takes place within a cell.
7 Respiratory zone	**g** The sense of smell.

Answers: 1. _____ 2. _____ 3. _____ 4. _____ 5. _____ 6. _____ 7. _____

2. The figure below shows the structures of the respiratory system. Complete the labelling of this diagram.

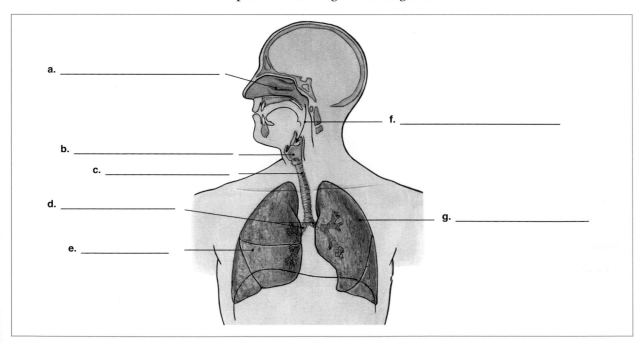

a. _____

f. _____

b. _____

c. _____

d. _____

g. _____

e. _____

3. Match the respiratory structures to their functions and write the answers in the spaces provided below. Please note that some structures have more than one function.

a. The organs of respiration

1. NOSE

f. Site of gaseous exchange

2. PHARYNX

b. Filters, warms and moistens air

3. LARYNX

g. Carry air into the alveoli

c. Transports air into the bronchi

4. TRACHEA

5. BRONCHI

h. Receives olfactory stimuli and acts as a resonating chamber for sound

d. A resonating chamber for sound

6. LUNGS

e. Produces sound

i. A passageway for air, food and drink

7. ALVEOLI

Answers: 1. _____ 2. _____ 3. _____ 4. _____ 5. _____ 6. _____ 7. _____

4. Below is an outline of the blood supply to the lungs. Fill in the missing words.

a. Pulmonary _____ bring _____ blood to the lungs. _____ blood is then transported from the lungs to the heart by the pulmonary _____ .

b. _____ arteries bring oxygenated blood to the lung tissue. Most of this blood is returned to the heart by the _____ veins. However, some of it drains into the _____ veins which transport it to the superior _____ _____ and then to the heart.

5. Are the following statements true or false? Write T or F in the spaces provided.

a. Pulmonary ventilation is dependent on the existence of a pressure gradient between the pressure inside the lungs and outside the lungs. ____

b. Expiration is the movement of air from the atmosphere into the lungs. ____

c. Inspiration occurs when the diaphragm and external intercostal muscles contract. ____

d. During normal, quiet breathing, expiration is a passive process that does not involve any muscular contraction. ____

e. In external respiration, oxygen moves from the alveolar air into the blood through the process of osmosis. ____

f. In internal respiration, blood loses carbon dioxide and gains oxygen. ____

g. Air is a mixture of oxygen, carbon dioxide, nitrogen, water vapour and a small quantity of inert gases. ____

Common pathologies of the respiratory system

6. Identify the following pathologies.

1. A chronic disorder characterised by 'attacks' in which a person struggles to exhale, has difficulty breathing and wheezes and coughs. Attacks can be caused by stimuli ranging from pollen and house dust mites to cold air or emotional upsets. _____

2. An irreversible disease in which the alveolar walls disintegrate, leaving abnormally large air spaces in the lungs. The main symptom of this disease is the inability to exhale easily. _____

3. Abnormally fast breathing when the body is at rest. Symptoms include dizziness, tingling sensations and a tightness across the chest. _____

4. Inflammation of the larynx. Characterised by hoarseness or voice loss, a sore throat and a painful or tickling cough. _____

5. Inflammation of the mucous membrane lining the nose. Symptoms include a runny or stuffy nose. _____

6. A contagious disease characterised by coughing, night sweats, decreased energy and decreased appetite. This disease is caused by a bacterium which is inhaled and usually affects the lungs by destroying parts of the lung tissue which is then replaced by fibrous connective tissue. _____

Vocabulary test

Match the words to their correct definition by matching the number to the letter. Write your answers in the spaces provided.

WORD	DEFINITION
1. Cough	**a.** The inefficient delivery of oxygen to the tissues.
2. Dyspnoea	**b.** A whistling sound produced when the airways are partially obstructed.
3. Hypoxia	**c.** A sudden, explosive movement of air rushing upwards through the respiratory passages.
4. Wheezing	**d.** Laboured or difficult breathing.

Answers: 1. _____ 2. _____ 3. _____ 4. _____

Multiple choice questions

1. **Which body system does the respiratory system work closely with to ensure there is a continuous supply of oxygen to all cells and a continuous removal of carbon dioxide?**
   a. Cardiovascular system
   b. Endocrine system
   c. Lymphatic system
   d. Urinary system.

2. **What is the correct term used to describe the exchange of gases?**
   a. Desquamation
   b. Oxidation
   c. Respiration
   d. Ventilation.

3. **When does blood gain oxygen and lose carbon dioxide?**
   a. During pulmonary ventilation.
   b. During external respiration.
   c. During internal respiration.
   d. During oxidation.

4. **When does blood lose oxygen and gain carbon dioxide?**
   a. During pulmonary ventilation.
   b. During external respiration.
   c. During internal respiration.
   d. During oxidation.

5. **Which of the following structures are all resonating chambers for sound?**
   a. Nasal cavity, paranasal sinuses and bronchi.
   b. Mouth, trachea, lungs, bronchi and larynx.
   c. Nasal vestibule, nasal cavity, alveoli and paranasal sinuses.
   d. Pharynx, mouth, nasal cavity and paranasal sinuses.

6. **What is another term for nostrils?**
   a. External nares
   b. Internal nares
   c. Nasal vestibules
   d. Nasal concha.

7. **Which of the following are all regions of the pharynx?**
   a. Conchae, nasopharynx, larynx.
   b. Oropharynx, sinuses, hypopharynx.
   c. Nasopharynx, oropharynx, laryngopharynx.
   d. Oropharynx, larynx, laryngopharynx.

8. **Which of the following best describes the structure of the larynx?**
   a. The larynx is a funnel-shaped tube whose walls are made up of skeletal muscles lined by mucous membrane and cilia.
   b. The larynx is a short passageway made up of eight pieces of rigid hyaline cartilage and a leaf-shaped piece of elastic cartilage.
   c. The larynx is a long, tubular passageway composed of 16–20 incomplete C-shaped rings of hyaline cartilage.
   d. The larynx is a short passageway composed of repeatedly dividing branches of incomplete rings of cartilage lined with a mucous membrane.

9. **Which of the following is incorrect?**
   a. The choanae are commonly called the nostrils.
   b. The pharynx is commonly called the throat.
   c. The larynx is commonly called the voicebox.
   d. The trachea is commonly called the windpipe.

10. **What is the main function of the trachea?**
    a. Routes air and food into their correct channels.
    b. Produces sound.
    c. Acts as a passageway for air, food and drink.
    d. Transports air.

11. **Which of the following correctly depicts the branching of the bronchi?**
    a. Secondary bronchi, tertiary bronchi, bronchioles, primary bronchi.
    b. Tertiary bronchi, secondary bronchi, primary bronchi, bronchioles.
    c. Bronchioles, primary bronchi, tertiary bronchi, secondary bronchi.
    d. Primary bronchi, secondary bronchi, tertiary bronchi, bronchioles.

12. **What is the name of the membrane that covers the lungs?**
    a. Parietal pleura
    b. Visceral pleura
    c. Parietal peritoneum
    d. Visceral peritoneum.

13. **Where does the gaseous exchange between the lungs and the blood occur?**
    a. At the pleura.
    b. At the mediastinum.
    c. At the respiratory membrane.
    d. At the bronchi.

14. **When does inspiration occur?**
    a. When the pressure within the lungs is lower than the pressure outside the lungs.
    b. When the pressure within the lungs is higher than the pressure outside the lungs.
    c. When the pressure within the lungs is equal to the pressure outside the lungs.
    d. When there is no pressure within the lungs.

15. **In external respiration, how does the exchange of gases take place?**
    a. Via simple diffusion.
    b. Via osmosis.
    c. Via facilitated diffusion.
    d. Via vesicular transport.

9 The Cardiovascular System

Revision/Self-study notes

Blood

1. Blood is a vital substance in the body that functions in transportation, regulation and protection. Describe each of these functions in more detail.

- Transportation _____
  _____
  _____

- Regulation _____
  _____
  _____

- Protection _____
  _____
  _____

2. Blood is composed of blood plasma, red blood cells and white blood cells. Complete the table below and on the next page.

COMPONENT	DESCRIPTION	FUNCTION
Blood plasma		
Erythrocytes		
Leucocytes		

COMPONENT	DESCRIPTION	FUNCTION
Leucocytes (granulocytes): *Neutrophils*		
Eosinophils		
Basophils		
Leucocytes (agranulocytes): *Lymphocytes*		
Monocytes		
Thrombocytes		

The heart

1. The figure below shows the structure of the heart. This diagram has been labelled for you, but next to each label write two or three words to describe (or explain to yourself) that structure. For example, next to the label *superior vena cava* you could write 'deoxygenated blood from above diaphragm'.

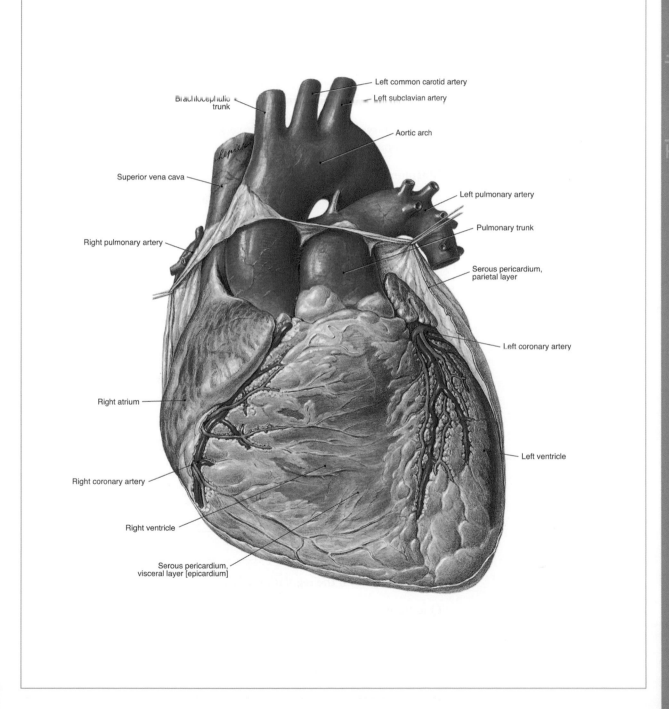

2. **The heart has four valves which prevent blood from flowing backwards. In the spaces provided below create your own mnemonic to remind you of each valve and its function.**

- **Atrioventricular valves**: lie between the atria and the ventricles and prevent blood from flowing back into the atria when the ventricles contract:

  – Tricuspid valve – right AV valve.

  Mnemonic:_____

  – Bicuspid valve – left AV valve.

  Mnemonic:_____

- **Semilunar valves**: lie between the ventricles and the arteries and prevent blood from flowing back into the ventricles when they relax:

  – Pulmonary semilunar valve – between the right ventricle and the pulmonary trunk.

  Mnemonic:_____

  – Aortic semilunar valve – between the left ventricle and the aorta.

  Mnemonic:_____

Physiology of the heart

1. **To ensure that you really understand what makes the heart beat and what controls or affects the heartbeat, put yourself in the examiner's seat and write questions on the topics given below. Then put yourself back in your own seat and answer the questions.**

Studytip

You may find it easier and more interesting if you do this exercise as part of a group.

a. **Ask a question on the intrinsic conduction system (nodal system):**

  Question: _____

  Answer: _____

  _____

  _____

b. **Ask a question on the regulation of the heart rate:**

  Question: _____

  Answer: _____

  _____

  _____

c. Ask a question on the cardiac cycle:

Question: _____

Answer: _____

_____

_____

d. Ask a question about blood pressure:

Question: _____

Answer: _____

_____

_____

Blood vessels

1. The cardiovascular system is composed of a closed system of tubes that all connect with one another to transport blood around the body. Complete the table below on the different types of blood vessels.

VESSEL	STRUCTURE	TRANSPORTS BLOOD FROM	TRANSPORTS BLOOD TO
Arteries			
Arterioles			
Capillaries			
Veins			
Venules			

2. The heart is a double-pump which pumps blood into two different circulations: pulmonary circulation and systemic circulation. In addition to this, there are two other recognised circulations: coronary circulation and hepatic portal circulation.

In the spaces provided on the next two pages, draw a spider diagram or mind map of each of these circulations.

Pulmonary circulation

Systemic circulation

Coronary circulation

Hepatic portal circulation

3. Now it's time to use your imagination! In order to learn the names of the many blood vessels in the body, picture it as a map and think of all the vessels as roads. A friend of yours is lost and needs directions so, giving each vessel a road name, direct your friend around the body.

Studytip

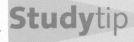

You may find it easier and more interesting if you do this exercise as part of a group. Have fun: create highways, roads, avenues, lanes and maybe even footpaths for the capillaries. You can either write the directions or draw them if you prefer.

To begin, take the Aorta Highway north out of the heart and turn…

Common pathologies of the cardiovascular system

Write down a short phrase that will remind you of each of the following
diseases and disorders.

Anaemia
- Iron-deficiency anaemia _____
- Pernicious anaemia _____
- Sickle-cell anaemia_____

Blood clotting disorders
- Haemophilia_____
- Thrombophilia_____
- Thrombosis_____
- Pulmonary embolism_____
- Epitaxis (nosebleeds) _____

Heart and blood vessel disorders
- Angina pectoris _____
- Arrhythmia_____
- Arteriosclerosis_____
- Atherosclerosis_____
- Coronary artery disease _____
- Gangrene_____
- Haemorrhoids (piles)_____
- Hypertension _____
- Hypotension_____
- Myocardial infarction (heart attack)_____
- Oedema_____
- Palpitations and panic attacks _____
- Phlebitis_____
- Raynaud's disease _____
- Varicose veins (varices) _____

Exercises

1. Complete the crossword below.

Across:

3. Organ that filters, cleans and stores blood (6).
4. What function does water perform in blood (7)?
6. The engulfment and digestion of foreign particles by white blood cells (12).
8. A white blood cell that plays an important role in the immune response (10).
9. A white blood cell (9).
14. Charged particles (ions) found in the blood. Examples include sodium, potassium, calcium and magnesium (12).
15. A protein that functions in blood clotting (10).
17. A watery, straw-coloured liquid that is a component of blood (6).

Down:

1. What does blood remove from cells (5)?
2. The oxygen-carrying molecule in blood (11).
5. A granulocyte that releases histamine (8).
7. A fluid connective tissue that is also called 'vascular tissue' (5).
10. A red blood cell (11).
11. The stopping of bleeding (11).
12. A white blood cell that engulfs and digests foreign particles and removes waste (10).
13. A chemical messenger transported by blood (7).
16. Gas that is carried by red blood cells (6).

2. Fill in the missing words.

a. The heart is a hollow, muscular organ divided into two halves. Both halves receive and deliver blood. The receiving chamber is called an _____ and the delivering chamber is called a _____ .

b. The _____ side of the heart receives _____ blood from the body and pumps it to the _____ .

c. The _____ side of the heart receives _____ blood from the lungs and pumps it to the _____ .

d. Separating the chambers of the heart are valves. Their main function is to prevent blood from flowing _____ . These valves are the _____ valves and the _____ valves.

3. The table below summarises the principal blood vessels of the heart. Complete the table.

VESSEL	BLOOD	FROM	TO
Aorta	a.	b.	Most of the body
Coronary artery	Oxygenated	c.	d.
Pulmonary artery	e.	Heart	f.
Superior vena cava	Deoxygenated	g.	h.
Inferior vena cava	i.	j.	Heart
Coronary sinus	k.	Heart tissue	Heart
Pulmonary vein	l.	m.	Heart

4. **The figure below shows the flow of blood through the heart. Complete the labelling of this image and, using the colour key, colour in the flow of blood (i.e. use red to colour in the flow of oxygen rich blood and blue to colour in the flow of oxygen poor blood).**

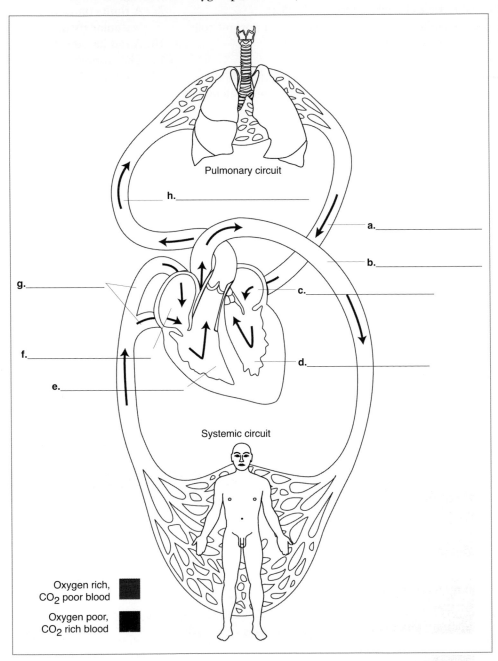

Pulmonary circuit

h._____

a._____

b._____

g._____

c._____

f._____

d._____

e._____

Systemic circuit

Oxygen rich, CO_2 poor blood ▮

Oxygen poor, CO_2 rich blood ▮

5. **Use the clues at the top of the next page to unscramble the letters below.**

a. ESORNITAALINDO

b. YOSSETL

c. YDRCAAICCCLE

_____

_____

_____

d. SATUHORTILHCMCLEY

e. LOERATREIS

f. LESIDAOT

_____

_____

_____

Clues:
- Specialised cells that form the pacemaker of the heart.
- A mass of autorhythmic cells found in the right atrial wall that initiate impulses and so start each heart beat.
- All the events associated with a heart beat.
- Contraction.
- Relaxation.
- Small vessels that carry blood away from the heart towards the tissues.

6. Dave is a young, adult male whose blood pressure is 160/100mm Hg.

a. Is Dave's blood pressure considered normal? If not, what is considered to be a normal blood pressure? _____

_____

b. Explain to Dave what the first figure, 160, and the second figure, 100, represent. _____

_____

7. Match the names of the blood vessels below to their descriptions and write the answers in the spaces provided.

Answers: 1. ____ 2. ____ 3. ____ 4. ____ 5. ____ 6. ____ 7. ____ 8. ____

a. Transports blood from the head, neck, upper limbs and thoracic wall to the heart	**1.** Great saphenous vein	**e.** Drain most of the blood from the lower limbs and pelvis and unite to form the inferior vena cava
b. Drains the inner leg and empties into the femoral vein	**2.** Renal arteries	
	3. Cephalic vein	**f.** Supply the kidneys with oxygenated blood
	4. Superior vena cava	
c. Drains the lateral aspect of the arm	**5.** Common iliac veins	**g.** Receives oxygenated blood from the left ventricle
	6. Hepatic portal vein	
d. Transports nutrient-rich, deoxygenated blood from the digestive organs to the liver	**7.** Aorta	**h.** Supply the toes with oxygenated blood
	8. Digital arteries	

Common pathologies of the cardiovascular system

8. Are the following statements true or false? Write T or F in the spaces provided.

1. Anaemia is an increase in the oxygen-carrying capacity of the blood. ____

2. Haemophilia is a hereditary disorder in which blood clots very slowly. ____

3. Deep vein thrombosis is the blocking of the pulmonary artery by an embolus. ____

4. The clinical term for nosebleeds is arrhythmia. ____

5. The clinical term for a heart attack is myocardial infarction. ____

6. Oedema is the excessive accumulation of interstitial fluid in body tissues. ____

7. Arteriosclerosis is the death and decay of tissue due to a lack of blood supply. ____

8. The common name for hypotension is high blood pressure. ____

9. Phlebitis is the inflammation of the walls of a vein. ____

10. Raynaud's disease is the temporary sensation of a chest pain caused by a lack of oxygen to the heart muscle. ____

Vocabulary test

Complete the table below.

WORD	DEFINITION
Diastole	a. _____
Systole	b. _____
Vasoconstriction	c. _____
Vasodilation	d. _____

Multiple choice questions

1. **What type of tissue is blood?**
   a. Epithelial
   b. Connective
   c. Muscle
   d. Nervous.

2. **Which of the following are all functions of blood?**
   a. Transportation, regulation, protection.
   b. Protection, locomotion, metabolism.
   c. Metabolism, excretion, thermogenesis.
   d. Excretion, locomotion, regulation.

3. **Which of the following are all types of white blood cells?**
   a. Erythrocytes, neutrophils, lymphocytes.
   b. Thrombocytes, plasma, haemoglobin.
   c. Basophils, eosinophils, monocytes.
   d. Fibrinogen, globulins, albumins.

4. **What is the name of the chamber of the heart that receives blood?**
   a. Atrium
   b. Ventricle
   c. Pericardium
   d. Apex.

5. **What is the name of the triple-layered sac that surrounds and protects the heart?**
   a. Pleura
   b. Perineum
   c. Peritoneum
   d. Pericardium.

6. **How many compartments, or chambers, is the heart divided into?**
   a. 1
   b. 2
   c. 3
   d. 4.

7. **Which valves lie between the atria and ventricles and prevent blood from flowing back into the atria when the ventricles contract?**
   a. Atrioventricular valves
   b. Semilunar valves
   c. Interventricular valves
   d. Bilunar valves.

8. **Which side of the heart receives deoxygenated blood from the body and pumps it to the lungs?**
   a. Right
   b. Left
   c. Both
   d. None of the above.

9. **What is the name of the vessels that carry blood away from the tissues towards the heart?**
   a. Arteries
   b. Arterioles
   c. Capillaries
   d. Veins.

10. **Which type of blood vessel has walls that are composed of a single layer of endothelium and a basement membrane?**
    a. Arteries
    b. Arterioles
    c. Capillaries
    d. Veins.

11. **Which of the following statements is true?**
    a. The structure of veins is the same as that of arteries.
    b. The walls of veins are thinner than those of arteries and some veins have valves to prevent the backflow of blood as it flows up towards the heart.
    c. The lumens of arteries are smaller than those of veins and some arteries have valves to prevent the backflow of blood as it flows away from the heart.
    d. Veins consist of a triple-layered wall surrounding a lumen while arteries consist of a double-layered wall surrounding a lumen.

12. **Which of the following statements best describes the flow of blood through vessels?**
    a. Blood flows from an area of high pressure into an area of low pressure.
    b. Blood flows from an area of low pressure into an area of high pressure.
    c. Blood flows from an area of high pressure into an area of equal pressure.
    d. Blood flows from an area of low pressure into an area of equal pressure.

13. **What is the name of the route blood follows from the heart to the tissues and organs of the body and back to the heart?**
    a. Coronary circulation
    b. Pulmonary circulation
    c. Renal circulation
    d. Systemic circulation.

14. **Which of the following blood vessels descend into the thighs to become the femoral arteries?**
    a. Popliteal arteries
    b. Popliteal veins
    c. Internal iliac arteries
    d. External iliac arteries.

15. **Which of the following blood vessels receives blood from the head, neck, upper limbs and thoracic wall?**
    a. Superior phrenic artery
    b. Inferior phrenic vein
    c. Superior vena cava
    d. Inferior vena cava.

10 The Lymphatic and Immune System

Revision/Self-study notes

The lymphatic system

1. The lymphatic system has three main functions. It drains interstitial fluid, transports dietary lipids and protects against invasions. Describe each of these functions in your own words.

• Drains interstitial fluid _____

_____

• Transports dietary lipids_____

_____

• Protects against invasions_____

_____

2. The lymphatic system is composed of a number of different vessels and structures through which lymph passes on its way to the subclavian veins. This journey has been outlined for you below. Create a spider diagram or mind map of the journey of lymph from interstitial fluid to its final destination.

Interstitial fluid → lymphatic capillaries → lymphatic vessels → lymphatic trunks → lymphatic ducts → subclavian veins.

Once you have traced this journey, add in the following structures and next to each structure describe its function.

• Lymphatic nodes • Thymus gland • Spleen • Lymphatic nodules.

Resistance to disease and immunity

Our bodies are constantly being attacked by a multitude of invaders and it defends itself in two different ways – through non-specific resistance to disease and through the immune response (immunity).

Studytip

You may find it easier and more interesting if you do this exercise as part of a group.

1. Imagine you are the general of an army that will attack a healthy person's body. You have different army personnel you can use, but you firstly need to inform them of what lies ahead. They know nothing about the body they are going to attack and you need to teach them about the different lines of defence they will encounter. On the next page is a list of what you need to include. In the box below, either write notes or sketch images that give details of each line of defence.

FIRST LINE OF DEFENCE	SPECIALISED DEFENCE
Mechanical barriers	Immunocompetent cells
Chemical barriers	Immune responses
Natural killer cells	Immunological memory
Phagocytes	
Inflammation	
Fever	

Studytip — You can make more of this exercise by taking time to discuss what could weaken a healthy body's defences. For example, poor nutrition, lack of exercise, stress, etc.

Common pathologies of the lymphatic and immune system

Write down a short phrase that will remind you of each of the following pathologies of the lymphatic and immune system.

- Allergy _____

- Lymphoedema _____

Cancer and the lymphatic and immune system

- Leukaemia _____

- Lymphoma_____

- Hodgkin's disease _____

- Non-Hodgkin's lymphoma _____

Infectious diseases

- Acquired immunodeficiency syndrome (AIDS)_____

- Glandular fever (Infectious mononucleosis) _____

- Tonsillitis _____

Autoimmune diseases and disorders

- Systemic lupus erythematosus (SLE) _____

Exercises

1. List the three main functions of the lymphatic system.

a. _____

b. _____

c. _____

2. Are the following statements true or false? Write T or F in the spaces provided.

a. Lymph is a clear fluid derived from interstitial fluid. It contains protein molecules, lipid molecules and red blood cells. _____

b. Lymph is only found in lymphatic vessels. _____

c. Interstitial fluid is absorbed into lymphatic capillaries where it becomes lymph. These capillaries transport it into lymphatic vessels which carry it through a number of lymphatic nodes and then into larger vessels called lymphatic trunks. _____

d. Lymphatic trunks carry lymph into two main ducts: the thoracic duct receives lymph from the upper right side of the body and the right lymphatic duct receives lymph from the entire body below the ribs. _____

e. Lymphatic nodes can also be called lymphatic glands. _____

3. Match the lymphatic structures to their functions and write the answers in the spaces provided below.

a. Lymphatic nodes	**1.** Filters and stores blood, destroys old worn-out red blood cells and produces lymphocytes.
b. Thymus gland	**2.** Help protect the body from pathogens that have been inhaled, digested or have entered the body via external openings.
c. Spleen	**3.** Produces hormones that help the development and maturation of T cells.
d. Lymphatic nodules	**4.** Filter lymph and produce lymphocytes.

Answers: a. _____ b. _____ c. _____ d. _____

4. The wordsearch below contains the names of different lymphatic nodes.
   Use the locations given below to find these names.

K	D	S	U	P	R	A	T	R	O	C	H	L	E	A	R	A	S	H
G	J	R	O	F	D	B	D	R	I	Y	A	D	X	N	I	R	P	O
S	U	P	E	R	F	I	C	I	A	L	P	A	R	O	T	I	D	N
U	S	G	E	D	A	P	Z	G	X	N	J	K	T	C	K	G	V	Y
B	Y	Y	Q	E	T	I	L	L	I	L	G	D	S	C	E	F	G	F
M	D	M	A	S	T	O	I	D	L	F	Q	Y	E	I	Q	W	H	S
A	K	L	D	U	A	S	F	G	L	G	A	F	W	P	Y	T	H	M
N	V	W	H	B	F	D	G	E	A	Y	E	U	Q	I	I	H	E	C
D	L	Z	E	M	H	J	J	D	R	G	D	K	D	T	J	X	W	P
I	W	W	X	E	C	K	J	O	Y	F	Y	J	A	A	G	W	N	O
B	R	R	X	N	K	L	G	U	Y	D	H	K	D	L	T	E	M	P
U	F	H	H	T	U	I	L	E	O	C	O	L	I	C	S	J	S	L
L	M	O	U	A	Y	L	R	H	Y	D	G	F	D	B	C	A	O	I
A	G	D	I	L	G	I	C	E	R	V	I	C	A	L	B	S	I	T
R	O	Y	O	A	D	A	A	M	D	H	R	S	G	E	R	R	C	E
C	I	E	K	W	L	C	S	G	C	R	F	D	Y	Y	Y	J	S	A
D	U	G	M	S	N	T	D	H	I	N	G	U	I	N	A	L	V	L

Location of lymphatic node:

- In front of ears
- Behind ears
- Beneath mandible
- Beneath chin
- Base of skull
- Neck

- Armpit
- Elbow crease
- Abdomen (near the diaphragm)
- Abdomen
- Groin
- Behind the knee.

5. The body employs two different approaches to defend itself against invaders. The first approach is non-specific resistance to disease. The second is the immune response. Each of these approaches uses a variety of different mechanisms, all given below. Link these mechanisms to their approaches.

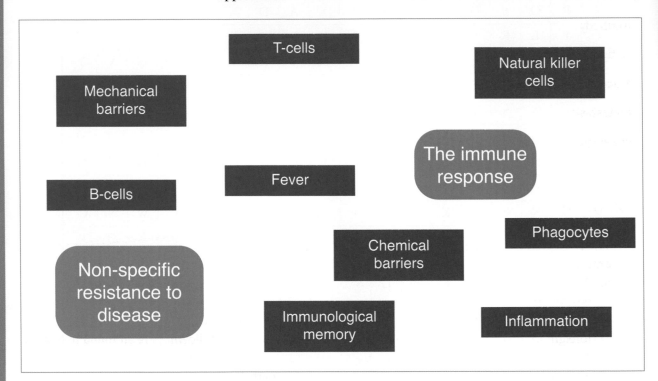

Common pathologies of the lymphatic and immune system

6. Name the following diseases and disorders.

1. Excessive swelling due to an accumulation of lymph in the tissues.

_____

2. Cancer of the white blood cells.

_____

3. A malignant lymphoma characterised by the progressive, painless enlargement of lymph nodes.

_____

4. A disease caused by the human immunodeficiency virus which results in a lowered T4 lymphocyte count and greater susceptibility to infections.

_____

5. Also called infectious mononucleosis, this disease is characterised by fatigue, headaches, dizziness and enlarged and tender lymph nodes.

_____

Vocabulary test

Complete the table below.

WORD	DEFINITION
Antibody	a. _____
Antigen	b. _____
Macrophage	c. _____
Metastasis	d. _____
Phagocyte	e. _____

Multiple choice questions

1. **Which of the following best describes an antigen?**
   a. A specialised protein that is synthesised to destroy foreign particles.
   b. A substance that the body recognises as foreign.
   c. A scavenger cell that engulfs and destroys foreign particles.
   d. A disease-causing micro-organism.

2. **Which of the following is not a function of the lymphatic system?**
   a. Immune response
   b. Drainage of interstitial fluid
   c. Thermogenesis
   d. Transportation of fats.

3. **Which of the following are transported by the lymphatic system?**
   a. Vitamins A,D,E and K
   b. Calcium, magnesium and zinc
   c. Carbohydrates
   d. Red blood cells.

4. **Where in the body do some lymphocytes become immune competent?**
   a. Liver
   b. Thymus
   c. Brain
   d. Kidneys.

5. **What is another name for lymphatic nodules?**
   a. Lymphatic vessels
   b. Lymphatic nodes
   c. Lymphatic trunks
   d. Mucosa-associated lymphoid tissue.

6. **Which of the following are all found in lymph?**
   a. Protein molecules, lipid molecules, foreign particles, cell debris, lymphocytes.
   b. Lipid molecules, bacteria, dead cells, erythrocytes.
   c. Platelets, bacteria, erythrocytes, dead cells.
   d. Foreign particles, cell debris, lymphocytes, erythrocytes.

7. **Where do the lymphatic ducts empty their contents?**
   a. Into lymphatic trunks.
   b. Into the left and right subclavian veins.
   c. Into lymphatic capillaries.
   d. Into the left and right carotid veins.

8. **Which of the following do not help lymph move through the lymphatic vessels?**
   a. Breathing movements
   b. Skeletal muscle
   c. B and T cells
   d. One-way valves.

9. **What is the function of lymphatic nodes?**
   a. To filter and clean blood.
   b. To protect the external openings of the body.
   c. To store platelets and blood.
   d. To filter lymph and produce lymphocytes.

10. **What structure can be described as follows: located on the left side of the body, this structure consists of white pulp and red pulp enclosed in a dense connective tissue from which trabeculae extend to form its framework?**
   a. Lymphatic nodes
   b. Lymphatic nodules
   c. Thymus
   d. Spleen.

11. **Where in the body are lymphatic nodules found?**
   a. Gastrointestinal tract and respiratory airways.
   b. Liver and large intestine.
   c. Heart and brain.
   d. Reproductive tract and gall bladder.

12. **What are the symptoms of inflammation?**
   a. Swelling and heat.
   b. Redness and pain.
   c. Occasionally the loss of function.
   d. All of the above.

13. **Which of the following are all mechanical barriers to disease?**
   a. Sebum and saliva.
   b. Lacrimal apparatus and gastric juice.
   c. The skin and urination.
   d. Perspiration and vaginal secretions.

14. **What type of cell are macrophages?**
   a. Erythrocytes
   b. Platelets
   c. Natural killer cells
   d. Phagocytes.

15. **What happens during an antibody-mediated immune response?**
   a. Killer cells attack antigens.
   b. Antibodies attack antigens.
   c. Phagocytes attack antigens.
   d. Macrophages attack antigens.

11 The Digestive System

Revision/Self-study notes

1. Describe the basic processes and functions of the digestive system.

* Ingestion_____

* Secretion_____

* Mixing and propulsion _____

* Mechanical digestion _____

* Chemical digestion _____

* Absorption_____

* Defecation _____

2. Describe the structure of the wall of the gastrointestinal tract.

_____

_____

_____

_____

_____

3. The digestive system may seem daunting. However, it is actually quite straightforward. You just need to learn the different structures of the system and what they do. Make some cue cards. On the front of the card write the name of the structure or organ and on the back write its function. Here is an example of a cue card for the mouth:

Mouth	**Mechanical digestion:** Mastication Deglutition **Chemical digestion:** Salivary amylase (carbohydrates) Lingual lipase (lipids)

Ensure you include these organs and structures of the gastrointestinal tract:
• Mouth • Pharynx • Oesophagus • Stomach • Small intestine (duodenum, jejunum and ileum) • Large intestine • Anus

And these accessory structures:
• Teeth, tongue, salivary glands • Liver • Gallbladder • Pancreas

Studygroups

To help you revise the digestive system use your cue cards to play the game 30 Seconds. Turn to page 10 to learn how to play.

4. Now that you know the basic structure and function of the digestive system, you need to learn about the digestion of carbohydrates, proteins and lipids. In the spaces below, draw a spider diagram or mind map to show the digestion of each of these.

Studytip

Be imaginative and have fun – the more you enjoy what you are doing the easier you will remember it. Use colour, symbols, images and even rhymes.

Digestion of carbohydrates

Digestion of proteins

Digestion of lipids

Common pathologies of the digestive system

Write down a short phrase that will remind you of each of the following diseases or disorders.

Pathologies of the mouth and teeth

• Apical abscess (tooth abscess) _____

• Gingivitis _____

• Halitosis _____

Pathologies of the oesophagus and stomach

• Dyspepsia (indigestion) _____

• Heartburn _____

• Hiatus hernia _____

• Hiccoughs _____

• Gastric ulcer _____

• Gastro-oesophageal reflux _____

Pathologies of the liver, gallbladder and pancreas

• Cholecystitis _____

• Cirrhosis of the liver _____

• Gallstones _____

• Hepatitis _____

• Hepatitis A _____

• Hepatitis B _____

• Hepatitis C _____

• Jaundice _____

Pathologies of the small and large intestines

• Appendicitis _____

• Coeliac disease _____

• Colitis _____

• Colon cancer (colorectal cancer) _____

• Constipation _____

- Crohn's disease _____

- Diarrhoea _____

- Diverticulosis _____

- Diverticulitis _____

- Enteritis _____

- Flatulence _____

- Gastritis _____

- Gastroenteritis _____

- Haemorrhoids (piles) _____

- Irritable bowel syndrome (IBS, spastic colon) _____

  _____

- Peptic ulcer _____

Eating disorders

- Anorexia nervosa _____

- Bulimia nervosa _____

- Obesity _____

Exercises

1. Complete the crossword on the following page.

Across

4. The pouch at the start of the large intestine that receives food from the small intestine (6).
5. The semifluid contents of the stomach (5).
6. Another term for chewing (11).
8. Another name for the large intestine (5).
10. The process by which indigestible substances are eliminated from the body (10).
12. The common term used for the buccal cavity (5).
13. Salivary glands below and in front of the ear (7).
16. The enzymes that digest fats (7).
17. A yellow, brownish liquid that is produced by the liver, stored in the gallbladder and necessary for the digestion of fats (4).
18. Another name for the gastrointestinal tract (10, 5)
20. Another term used for teeth (6).
21. The soft, flexible mass that food is reduced to after being chewed and mixed with saliva (5).

Down

1. Terminal opening of the gastrointestinal tract (4).
2. The act of swallowing (11).
3. The digestion of starches begins in the mouth when they come into contact with the enzyme salivary _____ (7).
7. The first segment of the small intestine (8).
9. The large serous membrane that lines the abdominal cavity (10).
11. The process of taking food into the mouth (9).
12. What molecules are carbohydrates broken down into during digestion (15).
13. An involuntary wave-like movement that pushes the contents of the gastrointestinal tract forward (11).
14. Functions of this vital organ include metabolism, detoxification and nutrient storage (5).
15. Long, tube running behind the trachea (10).
19. The name of the deepest layer of the wall of the gastrointestinal tract (6).

2. The columns below show the major digestive organs (orange boxes) and the enzymes/fluids that are found within them (blue boxes). However, some of these enzymes are in the wrong columns. Cross out those that are in the incorrect columns and write their names in the correct columns.

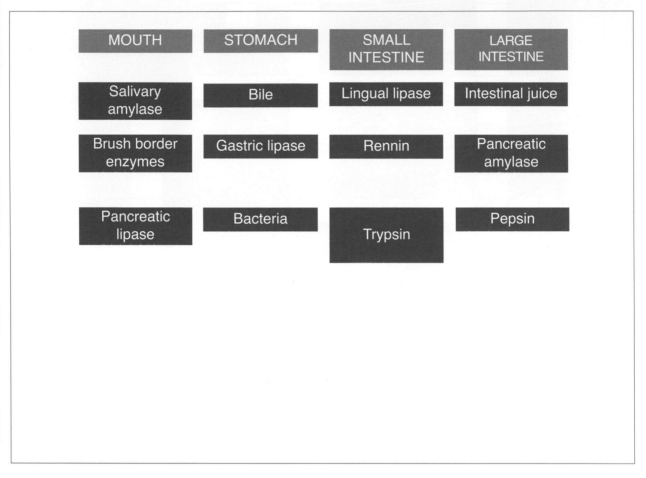

MOUTH	STOMACH	SMALL INTESTINE	LARGE INTESTINE
Salivary amylase	Bile	Lingual lipase	Intestinal juice
Brush border enzymes	Gastric lipase	Rennin	Pancreatic amylase
Pancreatic lipase	Bacteria	Trypsin	Pepsin

3. Are the following statements true or false? Write T or F in the spaces provided.

a. The digestion of carbohydrates begins in the mouth and is completed in the small intestine. _____

b. The digestion of proteins begins in the mouth and is completed in the stomach. _____

c. Maltase, sucrase and lactase are all enzymes that digest proteins. _____

d. Bile emulsifies fats in the small intestine. _____

e. Lipids are absorbed in the small intestine and transported to the liver via the hepatic portal vein. _____

4. The figure below is an overview of the organs and structures of the digestive system and their functions. However, the labels are incorrect. Relabel the diagram correctly by matching the label letter (a, b, c) to its correct line number (1, 2, 3). Write your answers in the spaces provided.

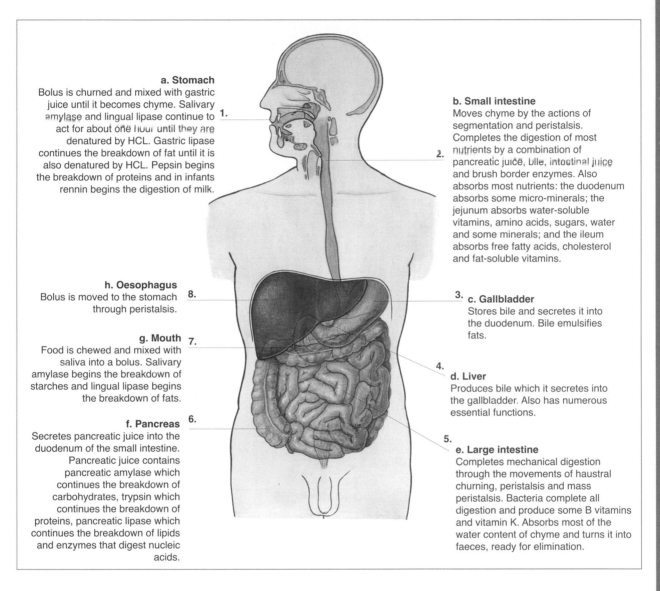

a. Stomach
Bolus is churned and mixed with gastric juice until it becomes chyme. Salivary amylase and lingual lipase continue to act for about one hour until they are denatured by HCL. Gastric lipase continues the breakdown of fat until it is also denatured by HCL. Pepsin begins the breakdown of proteins and in infants rennin begins the digestion of milk.

b. Small intestine
Moves chyme by the actions of segmentation and peristalsis. Completes the digestion of most nutrients by a combination of pancreatic juice, bile, intestinal juice and brush border enzymes. Also absorbs most nutrients: the duodenum absorbs some micro-minerals; the jejunum absorbs water-soluble vitamins, amino acids, sugars, water and some minerals; and the ileum absorbs free fatty acids, cholesterol and fat-soluble vitamins.

h. Oesophagus
Bolus is moved to the stomach through peristalsis.

c. Gallbladder
Stores bile and secretes it into the duodenum. Bile emulsifies fats.

g. Mouth
Food is chewed and mixed with saliva into a bolus. Salivary amylase begins the breakdown of starches and lingual lipase begins the breakdown of fats.

d. Liver
Produces bile which it secretes into the gallbladder. Also has numerous essential functions.

f. Pancreas
Secretes pancreatic juice into the duodenum of the small intestine. Pancreatic juice contains pancreatic amylase which continues the breakdown of carbohydrates, trypsin which continues the breakdown of proteins, pancreatic lipase which continues the breakdown of lipids and enzymes that digest nucleic acids.

e. Large intestine
Completes mechanical digestion through the movements of haustral churning, peristalsis and mass peristalsis. Bacteria complete all digestion and produce some B vitamins and vitamin K. Absorbs most of the water content of chyme and turns it into faeces, ready for elimination.

Answers: 1. _____ 2. _____ 3. _____ 4. _____ 5. _____ 6. _____ 7. _____ 8. _____

Common pathologies of the digestive system

5. In the table below match the disease or disorder to its correct description and write your answers in the spaces provided.

DISEASE/DISORDER		DESCRIPTION	
1	Bulimia nervosa	a	Inflammation of the gums.
2	Coeliac disease	b	Cancer of the large intestine.
3	Colorectal cancer	c	Psychological disorder characterised by bingeing and purging.
4	Crohn's disease	d	Condition characterised by recurring flare-ups of abdominal pain, constipation and diarrhoea in an otherwise healthy person.
5	Gingivitis	e	Yellowing of the skin or whites of the eyes due to high levels of bilirubin in the bloodstream.
6	Hepatitis	f	Chronic inflammation of the wall of the gastrointestinal tract.
7	Irritable bowel syndrome	g	Destruction of liver tissue and its replacement by scar tissue.
8	Jaundice	h	Intolerance to gluten.
9	Liver cirrhosis	i	A break in the mucous lining of the gastrointestinal tract usually due to the combined action of pepsin and hydrochloric acid.
10	Peptic ulcer	j	Inflammation of the liver.

Answers:

1. _____   6. _____

2. _____   7. _____

3. _____   8. _____

4. _____   9. _____

5. _____   10. _____

Vocabulary test

Complete the table below.

WORD	DEFINITION
Absorption	a. _____
Catalyst	b. _____
Digestion	c. _____
Substrate	d. _____

Multiple choice questions

1. The functions of the digestive system can be broken down into basic processes. What is the correct order in which these processes take place?
   a. Defecation, secretion, ingestion, absorption, digestion, propulsion, mixing.
   b. Ingestion, secretion, mixing, propulsion, digestion, absorption, defecation.
   c. Mixing, digestion, ingestion, secretion, defecation, propulsion, absorption.
   d. Absorption, ingestion, mixing, propulsion, digestion, secretion, defecation.

2. Which of the following is not part of the mechanical digestion of food?
   a. Mastication
   b. Deglutition
   c. Peristalsis
   d. Enzymatic action.

3. Which of the following are all part of the gastrointestinal tract?
   a. Mouth, pharynx, oesophagus, stomach, intestines.
   b. Mouth, teeth, tongue, oesophagus, intestines.
   c. Pharynx, gallbladder, liver, pancreas.
   d. Gallbladder, oesophagus, stomach, intestines.

4. What is the name of the membrane that lines the walls of the abdominopelvic cavity?
   a. Pericardium
   b. Pleural serous membrane
   c. Parietal peritoneum
   d. Visceral peritoneum.

5. What is another name for lips?
   a. Labia
   b. Uvula
   c. Frenulum
   d. Dentes.

6. Which of the following enzymes breaks down carbohydrates in the mouth?
   a. Lingual lipase
   b. Salivary amylase
   c. Pancreatic lipase
   d. Gastric amylase.

7. Where are gastric pits found?
   a. In the lining of the oesophagus.
   b. In the lining of the pharynx.
   c. In the lining of the stomach.
   d. In the lining of the small intestine.

8. What does the enzyme pepsin break down?
   a. Carbohydrates
   b. Lipids
   c. Proteins
   d. Fats.

9. Which enzyme is only found in the stomach of infants?
   a. Gastric lipase
   b. Pepsin
   c. Trypsin
   d. Rennin.

10. Which digestive organ is also considered an endocrine gland?
    a. Pancreas
    b. Liver
    c. Gallbladder
    d. Thymus.

11. Which of the following are all enzymes found in pancreatic juice?
    a. Trypsin, rennin, pepsin.
    b. Pancreatic amylase, trypsin, pancreatic lipase.
    c. Salivary amylase, lingual lipase, pepsin.
    d. Pancreatic amylase, pepsin, pancreatic lipase.

12. Which of the following is a function of the liver?
    a. Carbohydrate metabolism
    b. Protein metabolism
    c. Detoxification
    d. All of the above.

13. **Where is bile stored?**
    a. Liver
    b. Gallbladder
    c. Duodenum
    d. Bile ducts.

14. **Which of the following are all parts of the small intestine?**
    a. Ascending colon, jejunum, anus.
    b. Duodenum, jejunum, caecum.
    c. Duodenum, jejunum, ileum.
    d. Ascending colon, jejunum, rectum.

15. **Where does approximately 90% of all digestion take place?**
    a. Stomach
    b. Small intestine
    c. Large intestine
    d. Colon.

12 The Urinary System

Revision/Self-study notes

1. The kidney's function in cleaning the blood is well known, however, these vital organs play many other important roles in the body. The functions of the kidneys are listed below. Explain each of them in your own words.

- Regulation of blood composition and volume _____
  _____
  _____
  _____

- Regulation of blood pH _____
  _____
  _____
  _____

- Regulation of blood pressure _____
  _____
  _____
  _____

- Synthesis of calcitriol _____
  _____
  _____
  _____

- Secretion of erythropoietin _____
  _____
  _____
  _____

- Synthesis of glucose _____
  _____
  _____
  _____

2. **The figure below shows the internal structure of a kidney. Label the following structures and next to each label write that structure's function.**

- Fibrous capsule
- Renal cortex
- Renal medulla (renal pyramids)
- Renal column
- Renal pelvis
- Renal sinus
- Ureter
- Renal artery
- Renal vein.

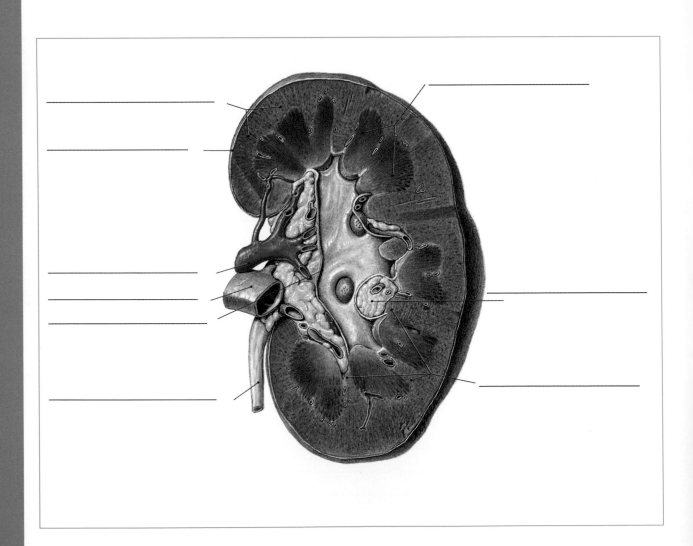

3. Nephrons are the functional units of the kidneys and it is here that urine is produced and that the volume and composition of blood is regulated. To help you understand how this happens, create a spider diagram or mind map that follows the route of blood through the nephron and the production of urine within the nephron. Ensure you include the following:

- **Renal corpuscle – Glomerular filtration:**
  - o glomerulus
  - o glomerular capsule
  - o afferent arteriole
  - o efferent arteriole
  - o filtration membrane.
- **Renal tubule – Tubular secretion and reabsorption:**
  - o proximal convoluted tubule
  - o loop of Henle
  - o distal convoluted tubule.

4. The kidneys excrete approximately 1–2 litres of urine a day. Complete the table below which lists the solutes present in the normal urine of a healthy person.

SOLUTE	DESCRIPTION
Urea	
Creatinine	
Uric acid	
Urobilinogen	
Inorganic ions	

5. There are a number of different hormones that are all associated with the functioning of the kidneys. Explain the role each of these hormones plays and beneath your explanation write a mnemonic or draw a symbol that will remind you of this hormone's function.

- Parathyroid hormone _____

    _____

    Mnemonic:_____
- Calcitonin _____

    _____

    Mnemonic:_____
- Aldosterone _____

    _____

    Mnemonic:_____
- Renin _____

    _____

    Mnemonic:_____
- Antidiuretic hormone (vasopressin) _____

    _____

    Mnemonic:_____

6. Once you have mastered the kidneys, the remaining structures of the urinary system are quite straightforward. However, it is easy to confuse the ureters with the urethra so write a short sentence describing each of the following.

- Ureters _____

  _____

- Bladder _____

  _____

- Urethra _____

  _____

Common pathologies of the urinary system

Write down a short phrase that will remind you of each of the following diseases and disorders.

- Calculi _____

- Enuresis _____

- Nephritis (Bright's disease) _____

  _____

- Nephroblastoma (Wilms' tumour) _____

  _____

- Pyelitis _____

- Renal failure _____

  _____

- Uraemia _____

- Urinary incontinence _____

  _____

- Urinary tract infection _____

  _____

- Pyelonephritis _____

- Ureteritis _____

- Urethritis _____

- Cystitis _____

Exercises

1. The kidneys have many vital functions in the body and some of these have been listed and explained in the table below. However, the explanations are muddled. Match the correct explanation to its function and write your answers in the spaces below.

FUNCTION	EXPLANATION OF FUNCTION
1 Regulation of blood composition and volume	**a** The kidneys help synthesise the active form of vitamin D.
2 Regulation of blood pH	**b** The kidneys secrete a hormone which stimulates the production of red blood cells.
3 Regulation of blood pressure	**c** During periods of starvation the kidneys function in gluconeogenesis.
4 Synthesis of calcitriol	**d** The kidneys filter out and excrete differing amounts of hydrogen ions from the blood.
5 Synthesis of erythropoietin	**e** The kidneys secrete an enzyme called renin which causes an increase in blood volume and pressure.
6 Synthesis of glucose	**f** The kidneys filter blood, remove from it any substances that are no longer needed and restore certain amounts of water and solutes to the blood as needed.

Answers: 1. ____ 2. ____ 3. ____ 4. ____ 5. ____ 6. ____

2. Fill in the missing words in the spaces below.

a. The kidneys are uniquely structured to filter _____ and produce _____ .

b. Above each kidney sits an _____ gland.

c. If you cut a kidney lengthwise you will see that it has three distinct regions: the outer renal _____ , the middle renal _____ and the inner renal _____ .

d. The functional units of the kidney, where urine is formed, are called _____ .

3. Once urine has formed in the nephron, it drains through a series of structures before it is excreted from the body. The flow chart below shows this flow of urine, but it is incorrect. Put it into the correct order.

Flow of urine

a. Nephron

b. Urethra

c. External environment

d. Minor calyx

e. Bladder

f. Papillary duct

g. Renal pelvis
⇓
h. Major calyx
⇓
i. Ureter

Answers:

1. _____

2. _____

3. _____

4. _____

5. _____

6. _____

7. _____

8. _____

9. _____

4. Nephrons filter blood through the process of glomerular filtration. The figure below shows a nephron with its associated blood vessels. Label this diagram.

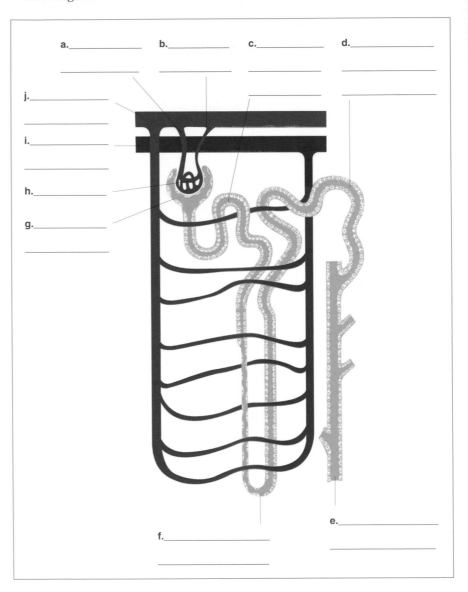

a._____   b._____   c._____   d._____

_____   _____   _____   _____

j._____

_____

i._____

_____

h._____

g._____

_____

f._____

_____

e._____

_____

5. Below is a list of substances that can be found in urine. Cross out all those which are not normally found in the urine of a healthy person.

Haemoglobin   Urobilinogen

Creatinine   Proteins

Pus   Red blood cells

Uric acid   Bile pigments

Inorganic ions   Urea

Glucose

6. Are the following statements true or false? Write T or F in the spaces provided.

a. Intracellular fluid is the fluid inside cells. _____

b. Blood plasma, cerebrospinal fluid, serous fluid and lymph are all examples of intracellular fluid. _____

c. Water functions as a solvent in the body. _____

d. The small tube leading from the bladder to the external environment is called a ureter. _____

e. Females have smaller bladders than males. _____

Common pathologies of the urinary system

7. In the table below match the disease or disorder to its correct description and write your answers in the spaces provided.

DISEASE/DISORDER	DESCRIPTION
1 Calculi	a Abnormal or unusual bed wetting.
2 Enuresis	b The kidney's inability to filter blood efficiently.
3 Nephritis	c The involuntary passing of urine.
4 Renal failure	d Inflammation of the kidneys.
5 Urinary incontinence	e Hard masses that form in the urinary tract. Commonly called kidney, bladder or ureteral stones.

Answers: 1. ____ 2. ____ 3. ____ 4. ____ 5. ____

Vocabulary test

Complete the table below.

WORD	DEFINITION
Diuretic	a. _____
Electrolyte	b. _____
Micturition	c. _____

Multiple choice questions

1. **Which of the following are all functions of the kidneys?**
   a. Thermogenesis, regulation of blood pressure, synthesis of erythropoietin.
   b. Regulation of blood composition and volume, regulation of blood pH, synthesis of calcitriol.
   c. Regulation of blood pressure, synthesis of glucose, desquamation.
   d. Regulation of blood pH, thermogenesis, deglutition.

2. **What is the name of the outer region of a kidney?**
   a. Renal cortex
   b. Renal medulla
   c. Renal pelvis
   d. Renal sinus.

3. **Where in a kidney is urine formed?**
   a. The hilux
   b. The calyx
   c. The nephron
   d. The pelvis.

4. **Where inside a nephron do the processes of secretion and reabsorption take place?**
   a. The glomerulus
   b. The afferent arteriole
   c. The renal corpuscle
   d. The renal tubule.

5. **What is the name of the closed end of the renal tubule?**
   a. The glomerulus
   b. The glomerular capsule
   c. The Loop of Henle
   d. The convoluted tubule.

6. **Where in the renal tubule does the reabsorption of the majority of substances occur?**
   a. Proximal convoluted tubule
   b. Loop of Henle
   c. Distal convoluted tubule
   d. Glomerulus.

7. **Where in the renal tubule does the 'fine-tuning' of the filtrate occur?**
   a. Proximal convoluted tubule
   b. Loop of Henle
   c. Distal convoluted tubule
   d. Glomerulus.

8. **What is the name of the vessels that bring blood to the kidneys?**
   a. Hepatic portal veins
   b. Hepatic arteries
   c. Renal veins
   d. Renal arteries.

9. **Which of the following statements is correct?**
   a. Efferent arterioles bring blood to the glomerulus and afferent arterioles drain blood from the glomerulus.
   b. Afferent arterioles bring blood to the glomerulus and efferent arterioles drain blood from the glomerulus.
   c. Peritubular arterioles bring blood to the glomerulus and peritubular venules drain blood from the glomerulus.
   d. Peritubular venules bring blood to the glomerulus and peritubular arterioles drain blood from the glomerulus.

10. **Which of the following solutes are all normal in the urine of a healthy person?**
    a. Urea, uric acid, inorganic ions.
    b. Glucose, proteins, red blood cells.
    c. Pus, proteins, urea.
    d. Haemoglobin, uric acid, red blood cells.

11. **Which of the following hormones all play a role in regulating tubular reabsorption?**
    a. Thyroid hormone, progesterone, calcitonin.
    b. Antidiuretic hormone, melanocyte-stimulating hormone, oestrogen.
    c. Parathyroid hormone, aldosterone, calcitonin.
    d. Antidiuretic hormone, prolactin, oestrogen.

12. **Which of the following is not an electrolyte?**
    a. Calcium
    b. Melatonin
    c. Potassium
    d. Sodium.

13. **What is a diuretic?**
    a. A substance that stimulates an increase in urine production.
    b. A substance that stimulates a decrease in urine production.
    c. A substance that stimulates an increase in blood pressure.
    d. A substance that stimulates a decrease in blood pressure.

14. **What is the name of the passageway that functions in discharging urine from the body?**
    a. The ureter
    b. The bladder
    c. The urethra
    d. The collecting tubule.

15. **If a person is suffering from fluid retention and blood in the urine which of the following disorders could they have?**
    a. Incontinence
    b. Nephritis
    c. Enuresis
    d. Nocturia.

13 The Reproductive System

Revision/Self-study notes

1. Reproduction is the process by which a new member of a species is produced and reproductive cell division occurs through a process called meiosis. However, meiosis is often confused with mitosis so, putting yourself into the examiner's position, make up four questions on these processes. Once you have written down these questions, see if you can answer them.

- Question 1: _____

    Answer: _____

    _____

    _____

    _____

- Question 2: _____

    Answer: _____

    _____

    _____

    _____

- Question 3: _____

    Answer: _____

    _____

    _____

    _____

- Question 4: _____

    Answer: _____

    _____

    _____

    _____

2. **The figure below is an overview of the male urinary and genital systems. Label the following structures and next to each label write a few words describing this structure's function.**

- Scrotum
- Testes
- Epididymis
- Vas deferens
- Urethra

- Seminal vesicles (glands)
- Prostate gland
- Bulbourethral glands
- Penis

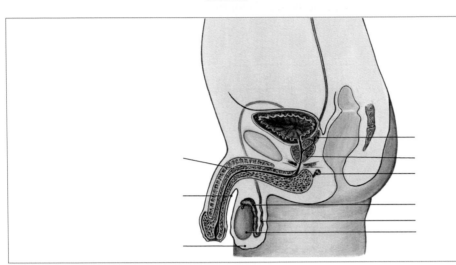

3. **The figure below is an overview of the female urinary and genital systems. Label the following structures and next to each label write a few words describing this structure's function.**

- Ovaries
- Fallopian tubes
- Uterus
- Vagina

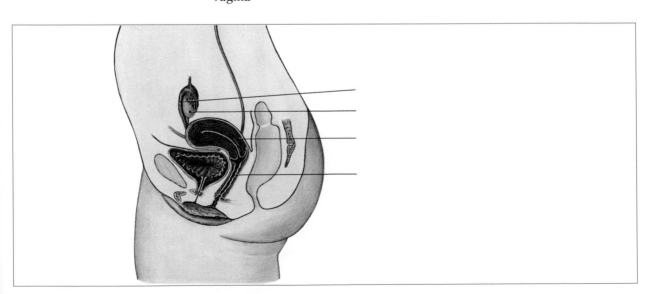

4. **To help you understand the female reproductive cycle, draw a spider diagram or mind map that describes the following phases. The first phase has been started for you.**

- Menstrual phase
- Preovulatory phase
- Ovulation
- Postovulatory phase

Stratum functionalis of endometrium dies and is discharged from body as menstrual flow.

Woman not pregnant, therefore...

Menstrual phase
Days 1–5

Hormones: oestrogen and progesterone levels decline

Studytip

Use colour and images to make your spider diagram or mind map as personal, fun and interesting as possible. That way you will find it easier to remember the facts!

5. Use a few memorable words or a short phrase to describe each of the following phases:

- Puberty...

- Pregnancy...

- Menopause...

Common pathologies of the reproductive system

Write down a short phrase that will remind you of each of the following diseases and disorders.

Diseases and disorders of the male reproductive system

- Benign prostatic hyperplasia (BPH) _____

  _____

- Impotence _____

- Prostate cancer _____

- Prostatitis _____

- Testicular cancer _____

Diseases and disorders of the female reproductive system

- Breast cancer _____

- Cervical cancer _____

- Ectopic pregnancy_____

- Endometriosis _____

- Fibroids _____

- Infertility _____

- Mastitis _____

- Amenorrhea_____

- Dysmenorrhea_____

- Menorrhagia _____

- Premenstrual syndrome _____

- Pelvic inflammatory disease (PID) _____

  _____

- Polycystic ovary syndrome (Stein-Leventhal syndrome) _____

  _____

  _____

• Postnatal depression _____

• Prolapsed uterus _____

• Toxic shock syndrome _____

• Vaginitis _____

Exercises

1. Define the words *meiosis* and *mitosis*.

a. Meiosis is _____

_____

b. Mitosis is _____

_____

2. Complete the table below which highlights the differences between meiosis and mitosis.

CHARACTERISTIC	MEIOSIS	MITOSIS
Number of daughter cells produced	a.	b.
Are daughter cells identical?	c.	d.
Number of chromosomes in each daughter cell	e.	f.

3. Connect the organs and structures below to where they belong.

Scrotum

Penis

Vulva

Ovaries

Vas deferens

Epididymis

Testes

Uterus

Male   Female

Vagina

Seminal vesicles

Spermatic cord

Bulbourethral glands

Prostate

Mammary glands

Fallopian tubes

4. Are the following statements true or false? Write T or F in the spaces provided.

a. The testes are the gonads of the female reproductive system. _____

b. The testes produce sperm through the process of spermatogenesis. _____

c. The testes also produce sex hormones called oestrogens which stimulate the development of masculine secondary sex characteristics. _____

d. The journey of sperm from the testes into the penis is as follows: formed in the seminiferous tubules of the testes → travels down the epididymis where they mature → into the vas deferens → then into the spermatic cord → the spermatic cord takes the sperm into the penis where it becomes semen. _____

e. The function of the penis is to excrete urine and ejaculate semen. _____

5. Use the clues given to unjumble the letters below.

AENSTNMIRUOT   TOECOY   PFAOEIUALSBTLN

a._____   b._____   c._____

EMIDNRTMOEU   TMPURUUCSOEL   LIAOAAIELGCNFRFL

d._____   e._____   f._____

Clues:
- An immature ovum.
- A large fluid-filled follicle that ruptures to expel an ovum during ovulation.
- After ovulation this empty follicle develops into a glandular structure that produces female hormones.
- Two thin tubes running from the ovaries to the uterus
- The tissue that lines the inside of the uterus.
- The process by which the lining of the uterus is shed and discarded.

6. Below you will find three descriptions of the female reproductive cycle. Which description is correct?

a. Menstruation normally lasts approximately 5 days and the last day of menstruation is termed Day 1 of a woman's cycle. Before it begins, a woman's uterus is not yet prepared to receive a fertilised ovum. If it does not receive a fertilised ovum, the levels of oestrogens and progesterone increase and the stratum functionalis dies and is discharged from the body via menstrual flow. After menstruation, a mature Graafian follicle forms and the endometrium proliferates until around day 3 of a 28-day cycle when the Graafian follicle ruptures to release an ovum into the pelvic cavity. This is called ovulation and is the time in which a woman can become pregnant. For approximately 3 days after ovulation the endometrium awaits the arrival of a fertilised ovum. The endometrium is now thickened, highly vascularised and secreting tissue fluid and glycogen. If fertilisation has occurred, a woman will become pregnant. If it has not occurred, menstruation and the reproductive cycle begin again.

b. Menstruation normally lasts approximately 5 days and the first day of menstruation is termed Day 1 of a woman's cycle. Before it begins, a woman's uterus is prepared to receive a fertilised ovum. If it does not receive a fertilised ovum, the levels of oestrogens and progesterone decline and the stratum functionalis dies and is discharged from the body via menstrual flow. After menstruation, a mature Graafian follicle forms and the endometrium proliferates until around day 14 of a 28-day cycle when the Graafian follicle ruptures to release an ovum into the pelvic cavity. This is called ovulation and is the time in which a woman can become pregnant. For approximately 14 days after ovulation the endometrium awaits the arrival of a fertilised ovum. The endometrium is now thickened, highly vascularised and secreting tissue fluid and glycogen. If fertilisation has occurred, a woman will become pregnant. If it has not occurred, menstruation and the reproductive cycle begin again.

c. Menstruation normally lasts approximately 10 days and the first day of menstruation is termed Day 1 of a woman's cycle. Before it begins, a woman's uterus is prepared to receive a fertilised ovum. If it does not receive a fertilised ovum, the levels of androgens and testosterone decline and the stratum functionalis dies and is discharged from the body via menstrual flow. After menstruation, a mature sperm forms and the endometrium proliferates until around day 14 of a 28-day cycle when the Graafian follicle ruptures to release a sperm into the pelvic cavity. This is called fertilisation and is the time in which a woman can become pregnant. For approximately 14 days after ovulation the endometrium awaits the arrival of a fertilised ovum. The endometrium is now thickened, highly vascularised and secreting tissue fluid and glycogen. If fertilisation has occurred, a woman will become pregnant. If it has not occurred, menstruation and the reproductive cycle begin again.

Answer: _____

Common pathologies of the reproductive system

7. Identify the diseases and disorders below.

1. Inflammation of the prostate gland. Symptoms include lower back pain, the urge to urinate frequently and burning or painful urination.

   _____

2. A pregnancy in which the foetus develops outside of the uterus.

   _____

3. Inflammation of the breast. _____

4. The absence of a menstrual period. _____

5. A hormonal disorder in which follicles fail to ovulate and instead collect as cysts over the ovaries. Symptoms can include acne, weight gain and irregular vaginal bleeding. _____

Vocabulary test

Complete the table below.

WORD	DEFINITION
a._____	The union and fusion of an ovum and a spermatozoa to form a zygote.
b._____	A cell with a single set of chromosomes.
c._____	The secretion of milk by the mammary glands.
d._____	The production of mature ova in the ovaries.
e._____	Fluid containing sperm and a mixture of fluids secreted by the reproductive glands.

Multiple choice questions

1. **What is the name given to the process in which a reproductive cell divides to produce four daughter cells each with only 23 chromosomes?**
    a. Mitosis
    b. Meiosis
    c. Oogenesis
    d. Spermatogenesis.

2. **Which of the following structures is found in both male and female bodies?**
    a. Scrotum
    b. Uterus
    c. Urethra
    d. Testes.

3. **What is the function of the testes?**
    a. Mitosis
    b. Meiosis
    c. Oogenesis
    d. Spermatogenesis.

4. **In which of the following structures do sperm mature?**
    a. Epididymis
    b. Vas deferens
    c. Spermatic cord
    d. Urethra.

5. **Which of the following structures all secrete fluids that combine with sperm to form semen?**
    a. Seminal vesicles, mammary glands, bulbourethral glands.
    b. Prostate gland, corpus luteum, Graafian follicle.
    c. Bulbourethral glands, prostate gland, seminal vesicles.
    d. Mammary glands, corpus luteum, corpus albicans.

6. **What is the name of the hormone responsible for the development of masculine secondary sexual characteristics?**
    a. Follicle-stimulating hormone
    b. Oestrogen
    c. Progesterone
    d. Testosterone.

7. **Where are the ovaries located?**
    a. The inferior portion of the pelvic cavity beneath the uterus.
    b. The middle portion of the pelvic cavity inside the uterus.
    c. The middle portion of the pelvic cavity below the uterus.
    d. The superior portion of the pelvic cavity on either side of the uterus.

8. **Which of the following statements is true?**
    a. An ovarian follicle enlarges to become a Graafian follicle which ruptures to expel an oocyte in the process of ovulation.
    b. An ovarian follicle enlarges to become the corpus luteum which ruptures to expel an oocyte in the process of ovulation.
    c. An ovarian follicle enlarges to become a Graafian follicle which ruptures to expel a mature ovum in the process of ovulation.
    d. An ovarian follicle enlarges to become the corpus luteum which ruptures to expel a mature ovum in the process of ovulation.

9. **Where in the female reproductive system do the sperm and ovum unite and fuse to form a zygote?**
    a. Urethra
    b. Fallopian tubes
    c. Uterus
    d. Ovaries.

10. **What is the name of the layer of tissue that lines the uterus and is shed during menstruation?**
    a. Stratum functionalis
    b. Stratum basalis
    c. Perimetrium
    d. Myometrium.

11. **What is the name of the principal hormone that stimulates the production of milk in the mammary glands?**
    a. Luteinizing hormone
    b. Relaxin
    c. Inhibin
    d. Prolactin.

12. **How long is an average pregnancy?**
    a. One trimester
    b. Two trimesters
    c. Three trimesters
    d. Four trimesters.

13. **What are the female gonads called?**
    a. Ovaries
    b. Oocytes
    c. Ova
    d. Gametes.

14. **What is the name of a specialist who deals with the female reproductive system?**
    a. Oncologist
    b. Gynaecologist
    c. Urologist
    d. Haemotologist.

15. **What is the name of the disorder in which endometrial tissue develops outside of the uterus?**
    a. Benign prostatic hyperplasia
    b. Endometriosis
    c. Fibroids
    d. Mastitis.

Mock exam paper 1

1. **A plane which divides the body into upper and lower portions is:**
   a. Transverse
   b. Sagittal
   c. Oblique
   d. Frontal (coronal)
2. **The stomach is found mainly in:**
   a. Right upper quadrant
   b. Left upper quadrant
   c. Right lower quadrant
   d. Left lower quadrant
3. **The integumentary system comprises:**
   a. Lymph glands, antibodies and spleen
   b. Stomach, spleen and liver
   c. Heart, lungs and blood
   d. Skin, hair and nails
4. **The dermis is made up of:**
   a. Epithelial tissues only
   b. Connective tissues only
   c. Epithelial and connective tissues
   d. Neither epithelial nor connective tissues
5. **Which cells are present in the stratum basale (germinativum)?**
   a. Melanocytes
   b. Merkel cells
   c. Langerhans cells
   d. All of the above
6. **Mature bone cells are known as:**
   a. Osteoblasts
   b. Adipocytes
   c. Osteocytes
   d. Fibroblasts
7. **There are _____ phalanges in the human body:**
   a. 14
   b. 28
   c. 56
   d. 60
8. **Visceral muscle is:**
   a. Found in the walls of the heart
   b. Under the control of the will
   c. Striped in appearance
   d. Found in the walls of hollow organs
9. **Fascicles are:**
   a. Groups of tendons
   b. Bundles of fascia
   c. Bundles of muscle fibres
   d. Actin filaments

10. **What is lacking when ATP is made anaerobically?**
    a. Calcium
    b. Oxygen
    c. Lactic acid
    d. ADP
11. **The cranial nerves are:**
    a. Always sensory
    b. Always motor
    c. May be mixed
    d. Are never mixed
12. **A nerve impulse is known as:**
    a. Synapse
    b. Action potential
    c. Node of Ranvier
    d. Neurotransmitter
13. **The outer layer of the meninges, continuous with the periosteum of the cranium, is the:**
    a. Dura mater
    b. Arachnoid mater
    c. Cranial mater
    d. Pia mater
14. **Hormones bind to receptors located on:**
    a. Brain cells
    b. All cells
    c. Target cells
    d. Gland cells
15. **Which of the following is a true statement regarding the endocrine system?**
    a. The adrenal cortex is essential to life
    b. The thyroid is essential to life
    c. The ovaries are essential to life
    d. The testes are essential to life
16. **Which of the following statements is correct?**
    a. Hypersecretion of insulin causes diabetes mellitus
    b. Hypersecretion of melatonin causes seasonal affective disorder
    c. Hyposecretion of human growth hormone causes acromegaly
    d. Hyposecretion of oxytocin causes sterility
17. **Cilia are:**
    a. Tiny glands
    b. Tiny hairs
    c. Tiny nerves
    d. Tiny blood vessels

18. **The epiglottis prevents:**
    a. Choking
    b. Asthma
    c. Colds
    d. Laryngitis
19. **The heart lies in the area of the thorax known as the:**
    a. Septum
    b. Epicardium
    c. Myocardium
    d. Mediastinum
20. **Which one of the following is a high pressure system?**
    a. Arterial
    b. Venous
    c. Pulmonary
    d. Capillary
21. **The digestive system consists of:**
    a. The alimentary canal and the gastrointestinal tract
    b. The liver, gall bladder and pancreas
    c. The gastrointestinal tract only
    d. The alimentary canal, liver, gall bladder and pancreas
22. **Terms which are associated with the urinary system are:**
    a. Nephrology
    b. Renal
    c. Both of the above
    d. Neither of the above
23. **Which hormone below is not associated with renal function?**
    a. Prolactin
    b. Aldosterone
    c. ADH (Vasopressin)
    d. Parathormone
24. **The uterus is situated:**
    a. On the posterior abdominal wall
    b. In front of the bladder and behind the rectum
    c. Behind the bladder and in front of the rectum
    d. On the lateral walls of the pelvis
25. **The journey the sperm takes in order to fertilise the ovum is (in the correct order):**
    a. Vagina, uterus, ovary, cervix
    b. Fallopian tube, uterus, cervix, vagina
    c. Uterus, vagina, cervix, ovary
    d. Vagina, cervix, uterus, Fallopian tube

Mock exam paper 2

1. **The thoracic cavity contains the:**
   a. Heart, brain, liver
   b. Heart, lungs, oesophagus
   c. Brain, liver, lungs
   d. Oesophagus, heart, liver

2. **The imaginary line down the centre of the body is the:**
   a. Proximal line
   b. Distal line
   c. Midline
   d. Frontline

3. **Hair is found all over the body except:**
   a. Eyelids, lips and tongue
   b. Lips, soles and palms
   c. Palms, soles and ears
   d. Palms, lips and eyelids

4. **Cells which have a sensory function in the epidermis are:**
   a. Merkel cells
   b. Keratinocytes
   c. Melanocytes
   d. Langerhans cells

5. **The process by which cartilage is replaced by bone is called:**
   a. Osteology
   b. Ossification
   c. Osteocytology
   d. Osteoporosis

6. **The function of a ligament is to:**
   a. Connect muscle to bone
   b. Connect tendon to muscle
   c. Connect bone to bone
   d. Connect muscle to muscle

7. **Muscles can:**
   a. Maintain posture
   b. Move substances within the body
   c. Generate heat
   d. All of the above

8. **Tendons and aponeuroses are formed from:**
   a. Epimysium
   b. Perimysium
   c. Endomysium
   d. All of the above

9. **The integrative function of the nervous system is to:**
   a. Detect stimuli from both inside and outside of the body
   b. Make decisions regarding information received
   c. Stimulate the endocrine system to act
   d. Enable effectors to maintain homeostasis

10. **There are _____ pairs of spinal nerves:**
    a. 8
    b. 12
    c. 24
    d. 31

11. **Endocrine glands are called 'ductless' glands because:**
    a. Hormones are secreted directly into the blood stream
    b. Their ducts connect directly with the blood stream
    c. The production of hormones is intermittent
    d. They produce hormones

12. **Endocrine glands which also have an exocrine function are the:**
    a. Pituitary
    b. Pancreas
    c. Sebaceous
    d. Sudoriferous

13. **The only organ in the body which can perform gaseous exchange of oxygen and carbon dioxide is:**
    a. The skin
    b. The lung
    c. The kidney
    d. The liver

14. **The process by which air is drawn in and out of the lungs is:**
    a. Ventilation
    b. Internal respiration
    c. Pulmonary respiration
    d. All of these

15. **Which list best describes the functions of the cardiovascular system?**
    a. Respiration, protection, transportation
    b. Excretion, defence, regulation of temperature
    c. Transportation, protection, excretion
    d. Regulation, transportation, protection

16. **Erythrocytes are shaped like biconcave discs because:**
    a. They do not have a nucleus
    b. Haemoglobin molecules are also shaped like this
    c. The shape ensures a large surface area
    d. The shape enables them to clump together to prevent blood loss

17. **The thin, smooth lining of the inside of the heart is the:**
    a. Endocardium
    b. Visceral pericardium
    c. Parietal pericardium
    d. Epicardium

18. **These structures vary in size from a pin head to an almond, they filter lymph and remove or destroy harmful substances:**
    a. Lymph vessels
    b. Lymph ducts
    c. Lymph nodes
    d. Lymph capillaries

19. **The two lymphatic ducts are:**
    a. Right and left lymphatic
    b. Right and left subclavian
    c. Thoracic and right lymphatic
    d. Thoracic and left lymphatic

20. **The finger-like projection hanging down from the soft palate is the:**
    a. Labia
    b. Frenulum
    c. Uvula
    d. Papilla

21. **The head of the pancreas lies:**
    a. Under the liver
    b. In the curve of the duodenum
    c. Between the ileum and the jejunum
    d. In the left lower quadrant

22. **_____ results from protein breakdown:**
    a. Amino acids
    b. Glucose
    c. Fatty acids
    d. Glycogen

23. **The kidneys are to be found:**
    a. Just below the waist
    b. Between T12 and L3 vertebrae
    c. Attached to the posterior abdominal wall
    d. All of the above

24. **Which of the following is true?**
    a. The seminal vesicles secrete an acidic fluid which helps to nourish the sperm
    b. Secretions from the prostate gland help to reduce the motility and viability of sperm
    c. Semen is the combined secretions of the prostate, seminal vesicles and Cowper's glands
    d. Semen is slightly acidic in order to neutralise the environment within the vagina

25. **Breast feeding requires:**
    a. Prolactin
    b. Oxytocin
    c. Both of these
    d. Neither of these

Mock exam paper 3

1. **The study of disease in the body is:**
   a. Physiology
   b. Cytology
   c. Histology
   d. Pathology

2. **Cervical refers to the:**
   a. Legs
   b. Neck
   c. Arms
   d. Back

3. **The skin plays an important part in raising body temperature by:**
   a. Vasoconstriction
   b. Sweating
   c. Vasodilation
   d. None of the above

4. **This layer of skin is waterproof, has cells with no nuclei and is most apparent on the palms of the hands and soles of the feet:**
   a. Stratum basale (germinativum)
   b. Stratum granulosum
   c. Stratum lucidum
   d. Stratum corneum

5. **Haversian systems can be found:**
   a. In the long bones only
   b. In compact bone only
   c. In cancellous (spongy) bone only
   d. In both compact and cancellous bone

6. **The tibia is commonly known as:**
   a. The breast bone
   b. The knee cap
   c. The thigh bone
   d. The shin bone

7. **An aponeurosis is:**
   a. A large cylindrical ligament
   b. A flat sheet-like tendon
   c. The plasma membrane of a muscle cell
   d. A contractile element of skeletal muscle

8. **This type of muscle has spindle-shaped cells with a single nucleus:**
   a. Smooth
   b. Cardiac
   c. Skeletal
   d. All types of muscle

9. **Which of the following is not true?**
   a. The overall function of the nervous system is to maintain homeostasis
   b. The nervous system works with the endocrine system to maintain homeostasis
   c. The nervous system relies upon hormones to effect change
   d. The endocrine system is regarded as a 'slower' system than the nervous system

10. **Digestion is under the control of which part of the nervous system?**
    a. Peripheral
    b. Central
    c. Somatic
    d. Autonomic

11. **The hormones produced by the endocrine system are regulated by:**
    a. Positive feedback
    b. Negative feedback
    c. Neither positive nor negative feedback
    d. Both negative and positive feedback

12. **Which of the structures below is not situated in the brain?**
    a. Parathyroids
    b. Hypothalamus
    c. Posterior pituitary
    d. Anterior pituitary

13. **Which of the following are functions of the respiratory system?**
    a. Olfaction
    b. Breathing
    c. Produces speech
    d. All of the above

14. **Tissue respiration is the same as:**
    a. Pulmonary respiration
    b. External respiration
    c. Internal respiration
    d. Cellular respiration

15. **Blood is a:**
    a. Muscular tissue
    b. Epithelial tissue
    c. Connective tissue
    d. Nervous tissue

16. **The wall of the left ventricle is much thicker than the right because:**
    a. The left ventricle is much smaller than the right
    b. Blood has to travel a much longer distance
    c. The pressure in the left side is lower than in the right
    d. All of the above

17. **Which of the following statements is true regarding the lymphatic system?**
    a. Lymph is derived from interstitial fluid and is only found in lymphatic vessels
    b. Lymph has a milky/creamy appearance and is produced by the digestive tract
    c. Lymph capillaries are similar to blood capillaries and allow exchange of nutrients
    d. Lymph is derived from the blood and is part of the blood plasma

18. **The lymphatic vessels are most like veins in structure because:**
    a. They both return their contents to the heart
    b. They both have one-way valves
    c. Their walls are both only one cell thick
    d. Their walls are both thick and muscular

19. **Digestion is:**
    a. The taking of food and drink into the body
    b. The process by which unwanted food is eliminated from the body
    c. The breaking down of large molecules into smaller ones
    d. The process by which food substances enter the bloodstream

20. **Which teeth are used for cutting into food?**
    a. Canines
    b. Incisors
    c. Premolars
    d. Molars

21. **Which of the following is not a true statement?**
    a. More than half the body's weight is made up of water
    b. The body can survive longer without water than without food
    c. Urine is mostly made up of water
    d. The size and shape of every cell is maintained by water

22. **The kidneys are protected by:**
    a. Adipose tissue
    b. Bones of the pelvis
    c. The vertebral column
    d. All of the above

23. **Examples of extracellular fluid are:**
    a. Lymph
    b. Interstitial fluid
    c. Cerebrospinal fluid
    d. All of the above

24. **Meiosis results in _____ daughter cells:**
    a. 4
    b. 2
    c. 2 daughter and 2 son cells
    d. 0

25. **The testes:**
    a. Lie in the scrotal sac
    b. Produce testosterone
    c. Produce sperm
    d. All of the above

Mock exam paper 4

1. **The brain and spinal cord are contained within which body cavity:**
   a. Ventral body cavity
   b. Abdominal cavity
   c. Dorsal cavity
   d. Pelvic

2. **The hand is _____ to the elbow:**
   a. Deep
   b. Distal
   c. Proximal
   d. Superifical

3. **Keratinocytes:**
   a. Make up 90% of epidermal cells
   b. Produce a pigment to colour the skin
   c. Found only in the stratum basale
   d. Arise from the bone marrow

4. **The stage of hair growth which lasts the longest is:**
   a. Telogen
   b. Catagen
   c. Anagen
   d. None of the above

5. **Which is the only moveable bone in the skull?**
   a. Zygomatic
   b. Mandible
   c. Nasal
   d. Maxilla

6. **Which of the following joints can permit movement?**
   a. Fibrous joints
   b. Coronal sutures
   c. Cartilaginous joints
   d. Synarthrotic joints

7. **Which is the heaviest muscle in the human body?**
   a. Latissimus dorsi
   b. Gluteus maximus
   c. Teres major
   d. Gastrocnemius

8. **The definition of adduct is:**
   a. To move towards the midline
   b. To move away from the midline
   c. To bring the soles of the feet together
   d. To make the soles of the feet face outwards

9. **Dendrites:**
   a. Receive information into the cell body
   b. Insulate the axon
   c. Transmit information away from the cell body
   d. Transmit information between sensory and motor neurones

10. **The part of the brain which interprets visual images is:**
    a. Frontal lobe
    b. Temporal lobe
    c. Parietal lobe
    d. Occipital lobe

11. **Which of these is not caused by abnormal secretion of thyroid hormones?**
    a. Cretinism
    b. Tetany
    c. Myxoedema
    d. Goitre

12. **A 'temporary' endocrine gland is the:**
    a. Stomach
    b. Placenta
    c. Gall bladder
    d. Urinary bladder

13. **The Eustachian tubes open into the:**
    a. Nasopharynx
    b. Larynx
    c. Oropharynx
    d. Laryngopharynx

14. **The left lung has _____ lobes and the right lung has _____ lobes.**
    a. 3 and 2
    b. 1 and 2
    c. 2 and 1
    d. 2 and 3

15. **Bradycardia is:**
    a. An irregular heart rhythm
    b. A heart rate over 100 beats per minute
    c. A heart rate below 60 beats per minute
    d. None of the above

16. **Which of the lists below best describes the functions of the lymphatic system?**
    a. Drainage, filtration, immunity, respiration
    b. Immunity, fat transport, production of lymphocytes, respiration
    c. Drainage, fat transport, immunity, filtration
    d. Filtration, immunity, production of lymphocytes, fat transport

17. **Inflammation is:**
    a. One of the body's responses to injury
    b. Characterised by heat, swelling, redness and pain
    c. A mechanism to prevent the spread of further tissue damage
    d. All of the above

18. **Which of the following statements is false?**
    a. The peritoneum, unlike other serous membranes, has only one layer
    b. The peritoneum is the largest serous membrane in the body
    c. The peritoneum is composed of large folds
    d. One of the functions of the peritoneum is to bind organs to the abdominal cavity

19. **Obesity can be identified by assessing an individual's _____:**
    a. GIT
    b. HCL
    c. BMI
    d. DNA

20. **An individual who has intolerance to gluten suffers from _____ disease.**
    a. Coeliac
    b. Colitis
    c. Crohn's
    d. Colorectal

21. **Most reabsorption in the nephron occurs in the:**
    a. Bowman's capsule
    b. Collecting duct
    c. Proximal convoluted tubule
    d. All of the above

22. **The majority of water in the body is found:**
    a. In the blood
    b. Inside the cells
    c. In the digestive system
    d. Outside the cells

23. **The uterus is lined with:**
    a. Endometrium
    b. Perimetrium
    c. Peritoneum
    d. Myometrium

24. **Which one of the following is not a sexually transmitted disease?**
    a. Syphylis
    b. Vaginitis
    c. Chlamydia
    d. AIDS

25. **Severe postnatal depression is linked to:**
    a. Mothers under the age of 20
    b. Mothers over the age of 40
    c. Infection
    d. Hormonal changes

Mock exam paper 5

1. The heart is _____ to the lungs.
   a. Proximal
   b. Lateral
   c. Medial
   d. Distal

2. Which of the following is true regarding lymph capillaries:
   a. They are blind (closed)-ended
   b. They are more permeable than blood capillaries
   c. They are not found in the brain
   d. All of the above are true

3. Examples of effectors are:
   a. Special senses
   b. Cranial nerves
   c. Muscles and glands
   d. Brain and spinal cord

4. Cholecystitis is:
   a. Inflammation of the stomach
   b. Inflammation of the gall bladder
   c. Inflammation of the liver
   d. Inflammation of the pancreas

5. The endocrine system maintains homeostasis with the nervous system but:
   a. Its effects are not as immediate
   b. Its effects are longer lasting
   c. The systems control each other
   d. All of the above

6. The functional units of the kidney are the:
   a. Columns
   b. Calyces
   c. Nephrons
   d. Pyramids

7. Which of the following contain the protein haemoglobin?
   a. Leucocytes
   b. Erythrocytes
   c. Monocytes
   d. Thrombocytes

8. Which of the following is a false statement:
   a. A zygote contains 46 chromosomes
   b. A zygote develops into a new organism
   c. Zygotes continue to divide by meiosis
   d. Zygotes result from fertilisation

9. Which of the following is the odd one out?
   a. Occipital
   b. Temporal
   c. Parietal
   d. Lacrimal

10. The liver receives nutrient-rich, deoxygenated blood from the digestive organs via the:
    a. Hepatic vein
    b. Hepatic portal vein
    c. Gastric vein
    d. Inferior mesenteric vein

11. The effect of 'jet-lag' is attributed to disruption of the production of:
    a. Melatonin
    b. Calcitonin
    c. Oxytocin
    d. Thymosin

12. Which of the following statements is incorrect?
    a. Filtration of the blood occurs in the glomerulus
    b. Filtration of the blood results in water and solutes being forced into the renal tubule
    c. The closed end of the renal tubule is called the collecting duct
    d. Blood cells are not filtered into the renal tubule

13. Chemical digestion of carbohydrates starts in the:
    a. Stomach
    b. Small intestine
    c. Mouth
    d. Large intestine

14. What is meant by atherosclerosis?
    a. Abnormal constriction of the walls of arterioles in the fingers and toes
    b. Inflammation of the walls of the veins
    c. Enlarged and dilated veins in the walls of the rectum and anus
    d. A build up of fatty substances on the walls of the arteries

15. Pulmonary refers to:
    a. Blood
    b. Nose
    c. Lung
    d. Heart

16. A drug or substance which stimulates an increase in urine production is called:
    a. Antibiotic
    b. Analgesic
    c. Hypnotic
    d. Diuretic

17. The spermatic cord contains:
    a. Urethra, penis, foreskin
    b. Blood and lymphatic vessels, nerves, muscles
    c. Prostate gland, seminal vesicles, bulbourethral glands
    d. Broad ligament, Graafian follicle, corpus luteum

18. The femoral artery can be found in the region of the:
    a. Ankle
    b. Sole
    c. Thigh
    d. Knee

19. The nose is lined with:
    a. Cancellous (spongy) bone
    b. Ciliated mucous membrane
    c. Serous membrane
    d. Yellow elastic tissue

20. Anabolism (protein synthesis) is stimulated by:
    a. Androgens and oestrogen
    b. Androgens only
    c. Testosterone only
    d. Progesterone

21. A high fibre diet can help to prevent:
    a. Ulcers
    b. Jaundice
    c. Constipation
    d. Diarrhoea

22. Varicose veins:
    a. May be a contraindication to massage
    b. May be an inherited condition
    c. May be caused by inefficient valves
    d. All of the above

23. In external respiration the blood loses _____.
    a. Oxygen
    b. Nitrogen
    c. Nothing
    d. Carbon dioxide

24. Prolonged hyperventilation can lead to:
    a. Pneumonia
    b. Unconsciousness
    c. A runny nose
    d. All of the above

25. Which endocrine gland is also part of the immune system?
    a. Thyroid
    b. Adrenal cortex
    c. Thymus
    d. Adrenal medulla

26. Lipids and lipid soluble vitamins are absorbed into the:
    a. Liver
    b. Blood capillaries
    c. Lymph nodes
    d. Lacteals

27. The thenar eminence is found:
    a. On the forehead
    b. At the base of the big toe
    c. At the base of the thumb
    d. At the base of the spine

Mock exam paper 5 cont.

28. Cartilage, tendons, ligaments and bone are all examples of:
    a. Connective tissue
    b. Nervous tissue
    c. Muscle tissue
    d. Epithelial tissue

29. Apoptosis is:
    a. A form of mitosis
    b. The process during which cells die
    c. A form of meiosis
    d. The pathological destruction of cells

30. An itchy rash in a specific area resulting from direct contact with cosmetics, metals or household chemicals could be:
    a. Urticaria (nettle rash or hives)
    b. Eczema
    c. Contact dermatitis
    d. Folliculitis

31. Carbon dioxide is transported by blood mainly in the:
    a. White cells
    b. Haemoglobin
    c. Plasma
    d. Red cells

32. When a muscle contracts but does not shorten and no movement results it is:
    a. Isometric contraction
    b. Isotonic contraction
    c. Eccentric contraction
    d. Concentric contraction

33. Mastitis is:
    a. More common around the time of childbirth
    b. Relieved by massage
    c. Inflammation of the vagina
    d. A yeast infection

34. Helper T-cells aid the:
    a. Cell-mediated immune response
    b. Antibody-mediated immune response
    c. Neither immune response is aided by helper T-cells
    d. Both immune responses are aided by helper T-cells

35. Which bone supports the tongue and provides attachment for muscles of the neck and pharynx?
    a. Hamate
    b. Hyoid
    c. Humerus
    d. Hyaline

36. Which part of the nervous system is regarded as conserving energy?
    a. Central nervous system
    b. Peripheral nervous system
    c. Sympathetic nervous system
    d. Parasympathetic nervous system

37. The technical terms 'onyx' and 'unguim' refer to:
    a. The skin
    b. The hair
    c. The nail
    d. All of the above

38. The term visceral refers to:
    a. The inner walls of a body cavity
    b. The outer surface of the body
    c. The internal organs of the body
    d. None of these

39. Anosmia is loss of the sense of:
    a. Smell
    b. Taste
    c. Sight
    d. Hearing

40. A basal cell carcinoma may commonly be known as:
    a. Melanoma
    b. Rodent ulcer
    c. Varicose ulcer
    d. Squamous cell carcinoma

41. Water makes up about _____ of our bodies.
    a. 10%
    b. 30%
    c. 60%
    d. 100%

42. If you push your head right backwards (whilst standing upright) to face up to the ceiling you are:
    a. Hypoextending the neck
    b. Rotating the neck
    c. Hyperextending the neck
    d. Circumducting the neck

43. Which is true of stress?
    a. Can be good for some people
    b. Can be alleviated by complementary therapies
    c. Can make people more susceptible to illness and disease
    d. All are true

44. The muscular dystrophies are:
    a. Infectious conditions
    b. Inherited conditions
    c. Caused by injuries
    d. Only occur in women

45. Saliva is _____ in nature.
    a. Acidic
    b. Alkaline
    c. Salty
    d. Semisolid

46. Renal failure:
    a. Is a life-threatening condition
    b. Could result in having dialysis
    c. Could result in having a kidney transplant
    d. All of the above

47. Which of the following is a false statement?
    a. Some skeletal muscles are not attached to bones
    b. Tendons attach skeletal muscles to bones
    c. Skeletal muscles are attached to both bones and joints
    d. Ligaments attach bones together at joints

48. The body system which contains the largest organ in the body is:
    a. Digestive system
    b. Nervous system
    c. Integumentary system
    d. Immune system

49. Cervical refers to:
    a. Neck
    b. Legs
    c. Arms
    d. Back

50. This muscle's origin includes the occipital bone (back of head) and most of the cervical and thoracic vertebrae. It has upper, middle and lower fibres:
    a. Latissimus dorsi
    b. Trapezius
    c. Deltoid
    d. Erector spinae

Mock exam paper 6

1. **Which of the following is incorrect with regard to meiosis and mitosis?**
   a. Meiosis only occurs in the gonads
   b. Mitosis is essential for growth and repair of cells
   c. Meiosis results in the haploid number
   d. Mitosis ensures genetic diversity

2. **Which of these is a false statement regarding the nervous system?**
   a. The right cerebral hemisphere controls the right side of the body
   b. The olfactory nerve is purely a sensory nerve
   c. The sciatic nerve is the longest nerve in the body
   d. The meninges cover both the brain and spinal cord

3. **Which of the following is not a solute carried by the blood plasma?**
   a. Electrolytes
   b. Nutrients
   c. Proteins
   d. pH

4. **Prolonged use of a wheelchair or bedrest will result in _____ of the leg muscles.**
   a. Hypertrophy
   b. Fatigue
   c. Atrophy
   d. Tonicity

5. **In internal respiration the blood gains _____.**
   a. Nutrients
   b. Nitrogen
   c. Carbon dioxide
   d. Oxygen

6. **How many teeth does a mature adult have?**
   a. 20
   b. 22
   c. 30
   d. 32

7. **The main function of the ovary is to:**
   a. House the growing foetus
   b. Produce the follicle stimulating hormone
   c. Produce mature ova
   d. Receive the sperm

8. **The four main taste sensations are:**
   a. Sweet, sour, bitter, salty
   b. Salty, sweet, spicy, stale
   c. Bitter, sweet, salty, spicy
   d. None of these as there are more than four

9. **Collagen and elastin give skin:**
   a. Extensibility and strength
   b. Nutrition and hydration
   c. Insulation and shock absorption
   d. Sensitivity and protection

10. **The function of mucous within the respiratory system is to:**
    a. Exchange gases
    b. Trap impurities
    c. Prevent dehydration
    d. Prevent bleeding

11. **Symptoms of premenstrual tension in women include:**
    a. Emotional upset
    b. Oedema and weight gain
    c. Fatigue and lethargy
    d. All of the above

12. **Which of the following is a false statement regarding deep vein thrombosis?**
    a. May be caused by inactivity, pregnancy, following surgery
    b. Usually forms in the deep veins of the leg
    c. Is a hereditary disease and runs in families
    d. May become mobile in the blood stream and cause a pulmonary embolus

13. **Muscle cells have many mitochondria because:**
    a. Mitochondria are the working units of muscles
    b. All the mitochondria are not always working
    c. Muscle cells use much energy
    d. Muscle cells require much protein

14. **Digestion does not take place in:**
    a. The mouth
    b. The oesophagus
    c. The stomach
    d. The duodenum

15. **The function of the spinal cord is:**
    a. To connect the brain to the rest of the body
    b. To transmit sensory information from the PNS to the brain
    c. To transmit motor instructions from the brain to the PNS
    d. All of the above

16. **Red blood cells do not contain a nucleus because:**
    a. They only live for 120 days
    b. Maximum space is required for oxygen transport
    c. They are not able to reproduce
    d. All of the above

17. **The alveoli are structured to ensure:**
    a. A huge surface area
    b. No friction occurs
    c. Collapsing does not occur
    d. All of the above

18. **Which statement is correct?**
    a. The penis and clitoris are both made of erectile tissue
    b. The perineum is present in both men and women
    c. The mons pubis protects the symphysis pubis
    d. They are all correct

19. **This vitamin is synthesised in the skin:**
    a. Vitamin C
    b. Vitamin D
    c. Vitamin E
    d. Vitamin K

20. **Which of the following is false?**
    a. Cancer sometimes spreads from one site in the body to another
    b. Cancer can be spread via the lymphatic system
    c. Cancer cannot be spread via the blood stream
    d. It is wise to consider carefully before treating a client who is suffering from cancer

21. **Which of the following is a false statement regarding meningitis?**
    a. It is inflammation of the meninges
    b. Symptoms can include a headache, stiff neck or a rash
    c. It can be a fatal disease
    d. The cause is always viral

22. **Which muscle flexes, abducts and laterally rotates the hip joint and flexes the knee joint?**
    a. Serratus anterior
    b. Soleus
    c. Sartorius
    d. Supinator

23. **Examples of steroid hormones are:**
    a. Sex hormones
    b. Oxytocin
    c. Insulin
    d. All of the above

24. **How many vertebrae are there in the thoracic (dorsal) spine?**
    a. 7
    b. 12
    c. 22
    d. 33

25. **If a client reports he has seen blood in his urine he should be advised to:**
    a. Drink more water
    b. Visit his GP
    c. Change his diet
    d. Book in for reflexology

Mock exam paper 6 cont.

26. The largest single mass of lymphatic tissue in the body is the:
    a. Liver
    b. Adenoids
    c. Tonsils
    d. Spleen

27. Which of the following statements is true regarding tuberculosis?
    a. It is a hospital acquired infection
    b. It no longer occurs in the United Kingdom
    c. It is caused by an airborne bacterium
    d. All of the above are true

28. The costal cartilages are located:
    a. At the hips
    b. At the neck
    c. At the knee
    d. At the ribs

29. Conception can be prevented by:
    a. Ligation (constriction) of the vas deferens
    b. Manipulation of the hormones by artificial means
    c. Physical barriers to prevent the sperm from reaching the egg
    d. All of the above

30. Which of the following is correct for the order of blood flow?
    a. Body, right side of heart, pulmonary artery, lungs, left side of heart, aorta
    b. Left side of heart, lungs, pulmonary artery, aorta, body, right side of heart
    c. Lungs, right side of heart, body, aorta, left side of heart, pulmonary artery
    d. Right side of heart, pulmonary artery, body, left side of heart, aorta, lungs

31. Which list below would best describe processes co-ordinated by the endocrine system?
    a. Homeostasis, negative feedback, positive feedback, hormone stimulation
    b. Reproduction, growth, metabolism, homeostasis
    c. Repair, replace, reproduce, restimulate
    d. Hyposecretion, hypersecretion, stimulation, inhibition

32. Which one of the following statements about chloasma is false?
    a. It can occur in pregnant women
    b. It usually affects the forehead, cheeks and around the lips
    c. It usually fades with time
    d. It is caused by an under production of melanin

33. _____ fever is a contagious disease of the lymphatic system common in children and young adults and often called the 'kissing' disease.
    a. Rheumatic
    b. Scarlet
    c. Glandular
    d. Yellow

34. Blood and mucous present in the stools may be a sign of:
    a. Colitis
    b. Gingivitis
    c. Cirrhosis
    d. Halitosis

35. Gliding joints are capable of:
    a. Side to side and back and forth
    b. Rotation and circumduction
    c. Adduction and abduction
    d. None of the above

36. The kidneys receive approximately _____ mls of blood per minute.
    a. 12
    b. 120
    c. 1200
    d. 12000

37. Which of the following is not an epithelial tissue?
    a. Cuboid
    b. Cartilage
    c. Columnar
    d. Ciliated

38. The lymphatic system is part of the:
    a. Digestive system
    b. Cardiovascular system
    c. Immune system
    d. Urinary system

39. Carbohydrate metabolism is affected by:
    a. Cortisol
    b. Progesterone
    c. ADH
    d. FSH

40. Water is the _____ of all body fluids.
    a. Solution
    b. Solute
    c. Concentrate
    d. Solvent

41. Which of the following statements about simple diffusion is false?
    a. A concentration gradient from high to low is involved
    b. The process continues until equilibrium is reached
    c. The rate of diffusion is affected by temperature
    d. A carrier is needed to help larger substances across the membrane

42. The menopause is characterised by:
    a. Inability to bear children
    b. Cessation of ovulation
    c. Unresponsiveness of the ovaries to FSH
    d. All of the above

43. A painful condition of the joints due to a build-up of uric acid, often found in middle-aged men is:
    a. Bunions
    b. Sprain
    c. Gout
    d. Rickets

44. Dialysis is:
    a. Passing large amounts of urine
    b. Removal of part of a kidney
    c. Mechanical cleansing of the blood
    d. A bacterial infection of the kidney

45. Which statement is false?
    a. Stressed individuals are more likely to suffer from irritable bowel syndrome
    b. 'Heartburn' is caused by reflux of acidic stomach contents
    c. Flatulence is bad breath
    d. Hepatitis is inflammation of the liver from a variety of causes

46. Which statement is untrue of hyaline cartilage:
    a. Attaches the ribs to the sternum
    b. Is strong and elastic and found in the external ear
    c. Covers the epiphyses to reduce friction
    d. Allows growth at the epiphyseal plate

47. The layer of hair where pigment granules are found is:
    a. Cuticle
    b. Matrix
    c. Cortex
    d. Medulla

48. The thorax is superior to:
    a. Abdomen
    b. Shoulders
    c. Head
    d. Mouth

49. Which is the correct order of the layers of the epidermis from the most superficial to the deepest?
    a. Stratum corneum, stratum germinativum, stratum spinosum, stratum lucidum, stratum granulosum
    b. Stratum germinativum, stratum spinosum, stratum lucidum, stratum granulosum, stratum corneum
    c. Stratum germinativum, stratum spinosum, stratum granulosum, stratum lucidum, stratum corneum
    d. Stratum corneum, stratum lucidum, stratum granulosum, stratum spinosum, stratum germinativum

50. The purpose of the myelin sheath is to:
    a. Enclose the brain and spinal cord
    b. Slow down the transmission of impulses
    c. Wrap around the cell bodies
    d. Speed up transmission of impulses

Mock exam paper 7

1. **Blood plasma makes up about _____ of blood:**
   a. 35%
   b. 45%
   c. 55%
   d. 65%

2. **The gonads are controlled by:**
   a. LH and FSH
   b. TSH
   c. GH
   d. MSH and ADH

3. **In an ectopic pregnancy:**
   a. The foetus develops in the Fallopian tube
   b. The foetus develops outside of the uterus
   c. The foetus does not develop at all
   d. None of the above

4. **Lymph transports escaped plasma proteins which cannot return to the blood because:**
   a. They are too large
   b. There are too many of them
   c. They are too small
   d. Plasma proteins don't escape from the blood

5. **The hormones associated with renal function are:**
   a. Urobilinogen and haemoglobin
   b. Urea and rennin
   c. Calcitonin and creatinine
   d. Erythropoietin and calcitriol

6. **The soft, downy hair found over most of the body surface is:**
   a. Lanugo
   b. Follicular
   c. Vellus
   d. Terminal

7. **Which of the following statements is false:**
   a. The role of neurones is to transmit impulses
   b. Neuroglia are smaller and more numerous than neurones
   c. Neuroglia insulate, protect and nurture neurones
   d. Both neurones and neuroglia divide by mitosis

8. **Which of the following is a false statement regarding skeletal muscle cells:**
   a. The sarcolemma is the muscle cell membrane
   b. Each muscle cell has a nucleus in the centre
   c. The cytoplasm is called the sarcoplasm
   d. A muscle cell is called a muscle fibre

9. **Which of the following are salivary glands?**
   a. Parotid
   b. Submandibular
   c. Sublingual
   d. All of the above

10. **Functions of the skeletal system include:**
    a. Protection, secretion, absorption, movement
    b. Muscle attachment, movement, shape, excretion
    c. Mineral homeostasis, movement, support, absorption
    d. Support, mineral homeostasis, protection, movement

11. **Which muscles below strengthen and stabilise the shoulder joint?**
    a. Diaphragm, internal intercostals, external intercostals
    b. Erector spinae, latissimus dorsi, trapezius
    c. Subscapularis, supraspinatus, infraspinatus, teres minor
    d. Teres major, quadratus lumborum, levator scapulae

12. **The back of the body is referred to as:**
    a. Posterior
    b. Parietal
    c. Proximal
    d. Peripheral

13. **The ions required for transmission of nerve impulses are:**
    a. Sodium and chloride
    b. Potassium and calcium
    c. Sodium and potassium
    d. Calcium and chloride

14. **Which of the following statements is correct?**
    a. Bile is essential for protein metabolism
    b. Bile salts function in carbohydrate breakdown
    c. Bile is essential for activation of vitamin D
    d. Bile is produced in the liver and stored in the gall bladder

15. **The talus is a bone in the:**
    a. Leg
    b. Arm
    c. Ankle
    d. Wrist

16. **Kidney stones can be caused by:**
    a. Hypersecretion of parathormone
    b. Hyposecretion of thymosin
    c. Hyposecretion of thyroxin
    d. None of the above

17. **The hormone which regulates the reabsorption of sodium ions into the blood and the secretion of potassium ions into the filtrate is:**
    a. Parathormone
    b. Vasopressin (antidiuretic hormone)
    c. Aldosterone
    d. Calcitriol

18. **The epidermis is:**
    a. The layer of the skin which provides insulation and storage for fat
    b. The layer of the skin which contains blood vessels, nerves and sweat glands
    c. Engaged in a constant process of filtration
    d. Engaged in a constant process of cell renewal

19. **The brain stem comprises:**
    a. Right and left cerebral hemispheres
    b. Thalamus and hypothalamus
    c. Medulla oblongata, pons and midbrain
    d. Cerebellum and cranial nerves

20. **Cardiac muscle tissue:**
    a. Is under voluntary control
    b. Is smooth and non-striated
    c. Has bundles of branching cells
    d. Is found in several hollow organs

21. **Which of the following is combined with oxygen in order to produce ATP?**
    a. Glycogen
    b. Carbon dioxide
    c. Glucose
    d. ADP

22. **The popliteal space can be found:**
    a. At the ankles
    b. Behind the knees
    c. Between the ribs
    d. Under the arms

23. **Cells which are able to engulf, ingest and destroy infection, waste and foreign particles are:**
    a. Macrophages
    b. Monocytes
    c. Phagocytes
    d. All of the above

24. **Which of the following statements is false?**
    a. The causes of prostate and testicular cancers are unknown
    b. Polycystic ovary syndrome is cancer of the ovaries
    c. Strong family history is a risk factor for breast cancer
    d. Cervical cancer is often called the 'quiet killer'

Mock exam paper 7 cont.

25. **The large intestine is found in:**
    a. The upper quadrants of the body
    b. The lower quadrants of the body
    c. All the quadrants of the body
    d. None of the above

26. **The pacemaker of the heart can be modified by:**
    a. Drugs
    b. The brain
    c. Hormones
    d. All of the above

27. **Which of the following statements is correct?**
    a. Eccrine glands are larger than apocrine glands
    b. Apocrine glands become active at puberty
    c. Eccrine glands are responsible for body odour
    d. Apocrine glands have a role to play in body temperature

28. **Both facial and body massage:**
    a. Increase the blood supply to muscle tissue
    b. Relieve muscle fatigue and tension
    c. Hasten the removal of waste products from muscle tissue
    d. All of the above

29. **Which of the following is a true statement?**
    a. The common name for the epiglottis is the throat
    b. The lungs are contained within the abdominal cavity
    c. The common name for the trachea is the windpipe
    d. The trachea connects the larynx to the bronchioles

30. **A ball and socket joint is capable of:**
    a. Flexion and extension
    b. Adduction and abduction
    c. Rotation and circumduction
    d. All of the above

31. **The majority of a cell is made up of:**
    a. Organelles
    b. Cytoplasm
    c. Mitochondria
    d. Nucleus

32. **Massage should be performed:**
    a. Towards the feet
    b. Towards the lymph nodes
    c. Away from the heart
    d. Away from the head

33. **The cause of hiatus hernia is:**
    a. Hereditary
    b. Trauma
    c. Bacterial
    d. Unknown

34. **Amenorrhoea can be caused by:**
    a. The menopause
    b. Pregnancy
    c. Anorexia nervosa
    d. All of the above

35. **The bladder is:**
    a. Situated in the pelvic cavity behind the symphysis pubis
    b. Attached to the posterior abdominal wall
    c. Behind the urethra
    d. In the abdominal cavity behind the intestines

36. **Which of the following is not true?**
    a. Sun exposure accelerates ageing of the skin
    b. Smoking, alcohol and caffeine affect the appearance of the skin
    c. Exposure to the sun causes less melanin to be produced
    d. Exposure to the sun can harm the skin but can also improve certain conditions

37. **Which of these is false regarding cerebrovascular accident (stroke)?**
    a. Speech is always affected
    b. Caused by disruption of blood supply to a part of the brain
    c. High blood pressure is a risk factor
    d. There can be sudden weakness on one side of the body

38. **Which is the correct statement below for inspired air?**
    a. 4.5% carbon dioxide and 16% oxygen
    b. 0.04% carbon dioxide and 21% oxygen
    c. 0.04% carbon dioxide and 16% oxygen
    d. 4.5% carbon dioxide and 21% oxygen

39. **Which of the following does not contain yellow elastic tissue?**
    a. Bronchial tubes
    b. Lung tissue
    c. Urinary bladder
    d. Aorta

40. **Gangrene:**
    a. May be a complication of diabetes mellitus
    b. May result from frostbite
    c. May result from severe arteriosclerosis
    d. All of the above

41. **The hormone which is the 'opposite' to insulin is:**
    a. Glucagon
    b. Glycogen
    c. Glucose
    d. Glycolysis

42. **The hormone responsible for the maintenance of pregnancy is:**
    a. Oxytocin
    b. Prolactin
    c. Progesterone
    d. Testosterone

43. **'Haustra' are to be found:**
    a. In the small intestine
    b. In the large intestine
    c. In the stomach
    d. In the liver

44. **Nephritis:**
    a. Can occur anywhere in the kidney
    b. Can be known as Bright's disease
    c. Is inflammation of the kidneys
    d. All of the above

45. **Synovial fluid cannot be found:**
    a. In diarthrotic joints
    b. In bursae
    c. Within the knee joint
    d. In a long bone

46. **Which of the following is not a true statement?**
    a. Joints between the bones of the cranium are fused at birth
    b. The 'soft spots' found on the cranium are called fontanelles
    c. The cranium is composed of eight bones
    d. Joints between the bones of the cranium are called sutures

47. **An increase in metabolism, fast heart rate, anxiety and heat intolerance are all symptoms of:**
    a. Myxoedema
    b. Seasonal affective disorder
    c. Pregnancy
    d. Thyrotoxicosis

48. **Phagocytosis means:**
    a. The process by which antibodies are produced
    b. The process by which infections are 'remembered' and destroyed
    c. The process by which foreign particles are engulfed and digested
    d. The process by which B and T lymphocytes become mature

49. **The cervix lies:**
    a. In the neck
    b. Between the vagina and uterus
    c. Between the ovary and Fallopian tube
    d. In the breast

50. **A fever:**
    a. Is an elevation of core temperature
    b. Intensifies the effects of the body's defences
    c. Inhibits the activity of some pathogens
    d. All of the above

Mock exam paper 8

1. **The Fallopian tubes are lined with:**
   a. Squamous epithelium
   b. Ciliated epithelium
   c. Cuboid epithelium
   d. Transitional epithelium

2. **Which part of the tooth acts as a barrier against acids?**
   a. Enamel
   b. Root
   c. Crown
   d. Neck

3. **Which of the following is not true?**
   a. The size of the vertebrae increase down to the lower back in order to support the weight of the body
   b. Although vertebrae differ in size and shape, they all share a similar structural pattern
   c. The discs found between the vertebrae form diarthrotic joints in order to facilitate free movement
   d. The lumbar vertebrae provide attachment for the large muscles of the back

4. **Which of these is not a function of cerebrospinal fluid?**
   a. Protects the CNS by fighting infection
   b. Acts as a shock absorber
   c. Medium for exchange of nutrients and waste
   d. Maintains constant pressure around the CNS

5. **Which of the following are all functions of the urinary system?**
   a. Filters waste, regulates blood flow, excretes hormones, maintains pH
   b. Filters blood, forms urine, maintains pH, regulates fluid and salt balance
   c. Maintains calcium balance, forms faeces, regulates fluids, excretes urine
   d. Regulates blood flow, filters lymph, assists mineral balance, forms urine

6. **Which of the following best describe the functions of the skin?**
   a. Heat regulation, secretion, excretion, sensation
   b. Protection, sensation, production of vitamin E, absorption
   c. Excretion, protection, heat regulation, circulation
   d. Absorption, production of vitamin D, respiration, protection

7. **Which gases are exchanged in the lungs?**
   a. Oxygen and nitrogen
   b. Oxygen and carbon dioxide
   c. Oxygen and carbon monoxide
   d. Oxygen and air

8. **Lymphocytes:**
   a. Engulf and destroy harmful microbes
   b. Produce antigens
   c. Selectively filter white cells and platelets
   d. Produce antibodies

9. **Which of these hormones are secreted by the posterior pituitary gland?**
   a. Melanocyte-stimulating hormone and Prolactin
   b. Follicle-stimulating hormone and Luteinizing hormone
   c. Growth hormone and Thyroid-stimulating hormone
   d. Oxytocin and Vasopressin

10. **Why is the right kidney slightly lower down than the left?**
    a. Because the right kidney is heavier than the left
    b. Because the left kidney is pushed upwards by the spleen
    c. Because the right kidney is pushed down by the liver
    d. Because there is less fat covering the right kidney

11. **Which of the following is correct?**
    a. A sagittal plane divides the body into unequal portions
    b. A frontal plane divides the body into anterior and posterior portions
    c. An oblique plane divides the body into equal portions
    d. A transverse plane divides the body vertically into right and left portions

12. **The part of the hair which can be seen above the skin surface is the:**
    a. Bulb
    b. Shaft
    c. Follicle
    d. Root

13. **Muscle fatigue is not caused by:**
    a. A lack of ATP
    b. A build up of lactic acid
    c. The inability of the muscle to contract
    d. Too much glycogen

14. **An exaggerated lumbar curve characterised by a sway back and lower back ache is called:**
    a. Scoliosis
    b. Osteoporosis
    c. Lordosis
    d. Kyphosis

15. **A blood vessel made up of a single layer of endothelium is:**
    a. An artery
    b. A capillary
    c. An arteriole
    d. A vein

16. **The purpose of the C-shaped cartilage in the trachea is to:**
    a. Allow the oesophagus to expand
    b. Prevent collapse of the trachea
    c. Both of these
    d. Neither of these

17. **Which of these hormones has an effect upon the skeletal system?**
    a. Calcitonin
    b. Parathormone
    c. Growth hormone
    d. All of the above

18. **Muscles which assist, and are found alongside, the prime mover are called:**
    a. Synergists
    b. Agonists
    c. Antagonists
    d. All of the above

19. **Hodgkin's disease is:**
    a. Cancer of red blood cells
    b. Cancer of the tonsils
    c. Cancer of the spleen
    d. Cancer of lymphocytes

20. **Ovulation usually occurs:**
    a. Mid cycle
    b. At the start of the cycle
    c. Towards the end of the cycle
    d. None of the above

21. **Which organ is essential to life?**
    a. The pancreas
    b. The stomach
    c. The gall bladder
    d. All of the above

22. **The function of the heart valves is:**
    a. To pump blood at a steady rate
    b. To ensure blood flow to the lungs
    c. To divide the heart into chambers
    d. To prevent backflow of blood

23. **The phrenic nerve is part of the _____ plexus.**
    a. Cervical
    b. Brachial
    c. Lumbar
    d. Sacral

24. **Chromosomes are made up of:**
    a. ATP
    b. DNA
    c. RNA
    d. CHO

25. **The renal pelvis is connected to the:**
    a. Ureter
    b. Bladder
    c. Urethra
    d. None of the above

Mock exam paper 8 cont.

26. Which one of these statements is false regarding angina pectoris?
    a. It is caused by insufficient blood supply to the heart muscle
    b. The pain of angina is not relieved by rest and relaxation
    c. It can be brought on by exercise or emotional stress
    d. The pain may spread from the chest into the arm or jaw

27. Which of the following statements is untrue regarding antigens?
    a. They may activate the immune response
    b. They may trigger an allergic reaction
    c. The can be transplanted organs such as a heart or kidney
    d. They are usually produced by the body's own cells

28. The adrenal cortex produces _____ groups of hormones.
    a. 2
    b. 3
    c. 7
    d. 9

29. Which of these glands produce oil to protect the skin and hair?
    a. Sebaceous
    b. Sweat
    c. Apocrine
    d. Eccrine

30. Which muscle below does not flex the forearm?
    a. Biceps brachii
    b. Biceps femoris
    c. Brachialis
    d. Brachioradialis

31. The walls of the alveoli are lined with _____ to aid diffusion.
    a. Cuboid epithelium
    b. Ciliated epithelium
    c. Squamous epithelium
    d. Stratified epithelium

32. An 'autoimmune' disease is:
    a. An inherited disease
    b. One which a person suffers from and no-one else catches it
    c. One where the body's defence system attacks its own tissues
    d. A disease which automatically cures itself after a period of time

33. Which of the following is an incorrect statement regarding the cardiac cycle?
    a. It can be a cause of high blood pressure
    b. It lasts approximately 0.8 seconds
    c. It comprises all the events associated with one heart beat
    d. It comprises both systole and diastole

34. The vulva:
    a. Is part of the vagina
    b. Houses the external genitalia in women
    c. Houses the external genitalia in men
    d. Is part of the scrotal sac

35. The large intestine consists of:
    a. Caecum, colon, rectum, jejunum
    b. Rectum, ascending colon, ileum, descending colon
    c. Colon, duodenum, jejunum, anal canal
    d. Caecum, rectum, anal canal, colon

36. If you stand in the anatomical position with your palms facing forwards and turn your palms towards your thighs so that they face backwards, you have just:
    a. Abducted your forearms
    b. Adducted your forearms
    c. Laterally rotated your forearms
    d. Medially rotated your forearms

37. The water soluble vitamins are:
    a. A & B
    b. B & C
    c. D & K
    d. E & C

38. The pelvic cavity does not contain:
    a. The gall bladder
    b. The reproductive organs
    c. The urinary bladder
    d. Part of the large intestine

39. Water and salts are reabsorbed in the:
    a. Glomerulus
    b. Loop of Henle
    c. Efferent arteriole
    d. All of the above

40. Tinea ungium is:
    a. Psoriasis of the nail
    b. Discolouration of the nail caused by viral infection
    c. Ringworm of the nail caused by fungal infection
    d. Bacterial infection of the nail fold

41. Dyspnoea is:
    a. Painful or difficult breathing
    b. No breathing
    c. Increased rate of breathing
    d. Very slow breathing

42. The butterfly-shaped endocrine gland wrapped around the trachea is the:
    a. Thymus
    b. Thyroid
    c. Thalamus
    d. Hypothalamus

43. If fertilisation occurs, which structure houses a foetus until birth?
    a. Fallopian tubes
    b. Ovaries
    c. Uterus
    d. Urethra

44. Hinge joints are capable of :
    a. Abduction and adduction
    b. Rotation and circumduction
    c. Flexion and extension
    d. None of the above

45. The taste buds are located on the tongue in elevations called _____.
    a. Papillomae
    b. Papillae
    c. Pustules
    d. Pyramids

46. Which of the following is false?
    a. The majority of people who suffer from anorexia nervosa are female
    b. Anorexia nervosa can lead to death
    c. Anorexia nervosa has a high incidence in countries where there is food shortage
    d. Amenorrhoea accompanies anorexia nervosa

47. A clouding over of the lens of the eye commonly found in older people is:
    a. A corneal ulcer
    b. A cataract
    c. A glaucoma
    d. Conjunctivitis

48. Which of the following best describes the characteristics of muscle tissue?
    a. Excitability, conductivity, enforceability, irritability
    b. Elasticity, extensibility, excitability, contractility
    c. Contractility, sensitivity, conductivity, extensibility
    d. Irritability, excitability, extensibility, enforceability

49. Which of the following is a by-product of cellular respiration?
    a. Glucose
    b. Lactic acid
    c. Carbon dioxide
    d. Oxygen

50. The correct order in the conduction system through the heart is:
    a. Atrioventricular node, bundle of His, sinoatrial node, Purkinje fibres
    b. Sinoatrial node, atrioventricular node, bundle of His, Purkinje fibres
    c. Sinoatrial node, Purkinje fibres, bundle of His, atrioventricular node
    d. Purkinje fibres, atrioventricular node, bundle of His, sinoatrial node

Mock exam paper 9

1. Efferent neurones conduct impulses:
   a. From CNS to muscles and glands
   b. From muscles and glands to CNS
   c. From PNS to CNS
   d. From sensory receptors to PNS

2. The phalanges are examples of:
   a. Short bones
   b. Long bones
   c. Flat bones
   d. Irregular bones

3. The urinary bladder has _____ muscle in its walls.
   a. Voluntary
   b. Cardiac
   c. Smooth
   d. Striated

4. Peyer's patches are found:
   a. In the small intestine
   b. In the spleen
   c. In the large intestine
   d. In the liver

5. The perineum is located:
   a. Between the toes
   b. Under the arms
   c. Behind the knees
   d. Between the legs

6. An average normal pregnancy lasts _____weeks.
   a. 20
   b. 30
   c. 40
   d. 50

7. Gases carried in the blood are:
   a. Oxygen
   b. Nitrogen
   c. Carbon dioxide
   d. All of the above

8. Which of these structures is not found within the conducting zone?
   a. Alveoli
   b. Bronchi
   c. Nose
   d. Pharynx

9. The link between the endocrine and nervous systems is considered to be the:
   a. Mid brain
   b. Medulla oblongata
   c. Hypothalamus
   d. Thalamus

10. The surface area of food to be digested is increased by the work of the:
    a. Tongue
    b. Teeth
    c. Saliva
    d. Uvula

11. Which of the statements below is true of fascia?
    a. Made of dense, irregular connective tissue
    b. Supports the nerves and blood vessels which support the muscles
    c. Separates muscles into different groups
    d. All are true

12. The correct term for shedding of skin cells is:
    a. Detoxification
    b. Desquamation
    c. Exfoliation
    d. Excretion

13. It is possible for the body to survive without the:
    a. Liver
    b. Red bone marrow
    c. Spleen
    d. None of the above

14. Which of these statements is true:
    a. Reproduction is possible before puberty
    b. Reproduction is possible after the menopause
    c. It is not possible for men to reproduce after the age of 55
    d. It is possible to reproduce after the menarche

15. Prolactin targets the:
    a. Uterus to contract
    b. Development of secondary sex characteristics
    c. Body clock
    d. Mammary glands to produce milk

16. Looking at a picture of the heart, what is the bottom left hand chamber called?
    a. Right atrium
    b. Right ventricle
    c. Left atrium
    d. Left ventricle

17. Thickening and scarring of the lung tissue is known as:
    a. Pulmonary fibrosis
    b. Sarcoidosis
    c. Sinusitis
    d. Pulmonary embolism

18. The semimembranosus is part of the _____?
    a. Adductors
    b. Quadriceps
    c. Hamstrings
    d. None of the above

19. The root, shaft and follicle are all parts of which structure of the integumentary system?
    a. Nail
    b. Hair
    c. Skin
    d. All of the above

20. Hearing a constant buzzing or other noises which other people cannot hear could be:
    a. Vertigo
    b. Otitis media
    c. Tinnitus
    d. Earache

21. The atlas and axis form what sort of joint?
    a. Hinge
    b. Pivot
    c. Ball & socket
    d. Condyloid

22. A body system not involved in excretion is:
    a. The integumentary system
    b. The nervous system
    c. The respiratory system
    d. The urinary system

23. The only vitamin which is water soluble and not fat soluble is:
    a. Vitamin C
    b. Vitamin D
    c. Vitamin E
    d. Vitamin K

24. The structures which increase surface area in the digestive system are the:
    a. Villi
    b. Plicae Circulares
    c. Microvilli
    d. All of the above

25. Amenorrhoea occurs:
    a. In pregnancy
    b. Before puberty
    c. During the menopause
    d. All of the above

26. Deoxygenated blood is transported back to the heart by the:
    a. Aorta
    b. Venae cavae
    c. Coronary veins
    d. Pulmonary veins

27. Lateral epicondylitis is:
    a. Housemaid's knee
    b. Golfer's elbow
    c. Tennis elbow
    d. None of the above

28. Pes planus affects:
    a. Cervical spine
    b. Arches of the feet
    c. Cartilage in the knee
    d. All the joints in the body

Mock exam paper 9 cont.

29. Which of the following statements is correct?
    a. More lymphatic vessels enter a lymph node than leave it
    b. The majority of lymph is collected by the right lymphatic duct
    c. Creamy coloured lymph drains into the cisterna chyli from the spleen
    d. All the above

30. The seminiferous tubules, sustentacular cells and Leydig cells are all to be found in:
    a. The ovaries
    b. The testes
    c. The breast
    d. The placenta

31. Which of the following conditions can children be vaccinated against?
    a. Whooping cough
    b. Asthma
    c. Asbestosis
    d. Sinusitis

32. The correct order of structures in the digestive system through which food passes is:
    a. Mouth, stomach, liver, large intestine, rectum
    b. Oesophagus, stomach, duodenum, pancreas, colon
    c. Mouth, oesophagus, stomach, small intestine, large intestine
    d. Oesophagus, ileum, pancreas, caecum, rectum

33. Which of the following statements is incorrect?
    a. The internal urethral sphincter is composed of smooth muscle fibres
    b. Emptying or voiding the bladder is known as micturition
    c. The walls of the bladder contain stretch receptors
    d. The internal urethral sphincter is under voluntary control

34. Which is true regarding sickle cell anaemia?
    a. The haemoglobin of sufferers bends the red blood cells into a crescent shape
    b. People who suffer from it have a high resistance to malaria
    c. It is a genetic disorder
    d. All of the above

35. Which of the muscles below lies deepest in the body?
    a. Transversus abdominis
    b. Rectus abdominis
    c. External obliques
    d. Internal obliques

36. Bacterial infection of the skin surrounding the nails is called:
    a. Leuconychia
    b. Anonychia
    c. Paronychia
    d. Koilonychia

37. Gigantism does not occur after puberty is complete because:
    a. Children have smaller pituitary glands
    b. The long bones of an adult have stopped lengthening
    c. The growth hormone is no longer secreted after puberty
    d. All of the above

38. Stimulation of the sympathetic nervous system would result in:
    a. Increased heart rate, increase in gastric juice, bronchodilation
    b. Increased heart rate, reduced peristalsis, dilated pupil
    c. Bronchodilation, constricted pupils, increased saliva
    d. Decreased saliva, bronchoconstriction, dilated pupils

39. Oxygen and carbon dioxide pass through the cell membrane by:
    a. Simple diffusion
    b. Vesicular transport
    c. Facilitated diffusion
    d. Active transport

40. Polycystic ovary syndrome may cause:
    a. Weight gain
    b. Chest and facial hair
    c. Infertility
    d. All of the above

41. Which statement is true of neurotransmitters?
    a. They are always excitatory
    b. They are always inhibitory
    c. They can also be a hormone
    d. They are stored in the synaptic cleft

42. Bursae are found:
    a. Between tendons and bones
    b. Between ligaments and bones
    c. Between muscles and bones
    d. All of the above

43. The renal tubule has a/an _____ lining.
    a. Epithelial
    b. Mucous membrane
    c. Serous membrane
    d. Muscular

44. Anaphylactic shock is:
    a. A disease-causing micro-organism
    b. An extremely severe allergic reaction
    c. A butterfly rash across the nose and cheeks
    d. An extremely painful sore throat

45. Cytokinesis is:
    a. The process whereby cells are able to move
    b. The process during which cells die
    c. The process by which the cytoplasm divides
    d. All of the above

46. Which protein molecule in red blood cells has an enormous affinity for oxygen?
    a. Prothrombin
    b. Globulin
    c. Albumin
    d. Haemoglobin

47. One of the following is a false statement:
    a. Jaundice is a symptom, not a sign of disease
    b. Jaundice causes yellowing of the skin
    c. Jaundice can occur temporarily in the newborn
    d. Jaundice is caused by high levels of bilirubin

48. Which of the statements below is not true of psoriasis?
    a. Is improved by exposure to the sun
    b. Shows raised red patches with silvery scales
    c. May run in families
    d. Is a viral skin condition

49. Which is the correct order of lymph flow within the lymphatic system?
    a. Capillaries, ducts, subclavian veins, vessels
    b. Capillaries, vessels, ducts, subclavian veins
    c. Subclavian veins, ducts, vessels, capillaries
    d. Vessels, subclavian veins, capillaries, ducts

50. _____ result from protein breakdown.
    a. Fatty acids
    b. Glucose
    c. Amino acids
    d. Glycogen.

Mock exam paper 10

1. **Which of the following is a false statement regarding muscle tone?**
   a. It is absent during sleep or unconsciousness.
   b. It maintains posture
   c. It is improved by exercise and massage
   d. It is when every muscle fibre is relaxed

2. **Of the cells in the blood 99% of them are:**
   a. White cells
   b. Red cells
   c. Platelets
   d. Antibodies

3. **What is the function of transitional epithelial tissue?**
   a. To protect underlying organs
   b. To move substances along a passageway
   c. To line organs which secrete substances
   d. To allow organs to distend or expand

4. **Difficulty in childbirth, lifting heavy objects, chronic coughing or constipation could result in:**
   a. Vaginitis
   b. Toxic shock syndrome
   c. Prolapse of the uterus
   d. Prostatitis

5. **The structure which drains lymph from the right shoulder is:**
   a. Tonsils
   b. Right lymphatic duct
   c. Thoracic duct
   d. Right subclavian vein

6. **Which of the following is a false statement?**
   a. The kidneys help in the production of white blood cells
   b. The kidneys play a very important part in homeostasis
   c. The kidneys help to regulate blood pressure
   d. The kidneys help the body to use vitamin D

7. **Exchange of gases takes place in the _____ zone.**
   a. Ventilatory
   b. Conducting
   c. Respiratory
   d. All of the above

8. **The 'master gland' of the body is the:**
   a. Pancreas
   b. Pituitary
   c. Pineal
   d. Parathyroid

9. **The skin protects the body by:**
   a. Forming a physical barrier against micro-organisms
   b. Tanning in response to exposure to ultraviolet radiation
   c. Exhibiting erythema and/or rashes in response to antigens
   d. All of the above

10. **Which of the following is a false statement regarding deglutition?**
    a. It interrupts breathing
    b. It is a mechanical process
    c. It takes approximately 2 seconds
    d. It is involuntary

11. **The brain, meninges and spinal cord are part of the:**
    a. Central nervous system
    b. Peripheral nervous system
    c. Sympathetic nervous system
    d. Parasympathetic nervous system

12. **An exaggerated thoracic curve, common in the elderly and which can be a result of osteoporosis is:**
    a. Scoliosis
    b. Kyphosis
    c. Lordosis
    d. Spondylosis

13. **Luteinizing hormone:**
    a. Stimulates ovulation
    b. Stimulates production of testosterone
    c. Stimulates production of oestrogen
    d. All of the above

14. **In general, disorders of the muscular system are accompanied by:**
    a. Pain
    b. Limitation of movement
    c. Inflammation
    d. All of the above

15. **Which of the following is not a protein?**
    a. Steroid
    b. Enzyme
    c. Antibody
    d. Hormone

16. **Poor sperm motility may cause:**
    a. Impotence
    b. Menorrhagia
    c. Infertility
    d. Prostatitis

17. **Which of the following is a false statement regarding the spleen?**
    a. Destroys old and worn out erythrocytes
    b. Filters lymph and blood
    c. Produces lymphocytes
    d. Stores platelets and blood

18. **'Goose pimples' are:**
    a. An attempt to lower body temperature
    b. Caused by contraction of the arrector pili muscle
    c. A bacterial infection of hair follicles
    d. Caused by blocked sebaceous glands

19. **The pectoral girdle consists of:**
    a. Ribs and vertebrae
    b. Radius and ulna
    c. Clavicle and scapula
    d. Sacrum and pubis

20. **The nasal cavity is lined with a _____ membrane.**
    a. Mucous
    b. Serous
    c. Synovial
    d. Glandular

21. **Parkinson's disease is characterised by:**
    a. Chronic infection
    b. Involuntary muscle tremors and rigidity
    c. Recurrent seizures
    d. Insomnia

22. **Which of the following conditions is not a type of repetitive strain injury?**
    a. Shin splints
    b. Housemaid's knee
    c. Fibromyalgia
    d. Tendonitis

23. **Which artery supplies the heart muscle itself with blood?**
    a. Coronary
    b. Carotid
    c. Cardiac
    d. Cephalic

24. **Blood and lymphatic vessels, nerves and the ureters enter and leave the kidney at the:**
    a. Pelvis
    b. Hilum
    c. Cortex
    d. Medulla

25. **Androgens are not responsible for:**
    a. Secondary sex characteristics in women
    b. Growth of pubic and axillary hair in both sexes
    c. The lowering of the voice
    d. Secondary sex characteristics in men

Mock exam paper 10 cont.

26. What sort of infection is athlete's foot?
   a. Bacterial
   b. Fungal
   c. Viral
   d. None of the above

27. The vast majority of absorption takes place in the:
   a. Stomach
   b. Colon
   c. Small intestine
   d. Liver

28. Which of the following is an incorrect statement?
   a. Simple cuboidal epithelium covers the surface of the ovaries
   b. Adipose tissue is found in the mammary glands
   c. Smooth muscle tissue is found in the myometrium
   d. Ciliated epithelial tissue lines the vagina

29. The presence of air in the pleural space is called:
   a. Pleurisy
   b. Pneumothorax
   c. Pneumonia
   d. None of the above

30. Which of the following muscles is named by its origin and insertion?
   a. Levator palpebrae superioris
   b. Deltoid
   c. Sternocleidomastoid
   d. Corrugator

31. DNA is the _____ of the cell.
   a. Genetic material
   b. Detoxification area
   c. Control centre
   d. Recycling centre

32. Onycholysis is:
   a. Nail biting
   b. Overgrowth of the cuticle
   c. Separation of the nail from its bed
   d. In-growing toenail

33. Which statement is false?
   a. The muscular layer of the stomach is called rugae
   b. The stomach can stretch and hold up to 4 litres
   c. The inner mucosal layer of the stomach contains gastric glands
   d. Some of the muscle fibres in the stomach are arranged obliquely

34. Which is true of the autonomic nervous system?
   a. It is under the control of the will
   b. It is an involuntary system
   c. It is part of the CNS
   d. It is controlled by the cerebellum

35. Which bone is butterfly-shaped and lies at the base of the skull?
   a. Temporal
   b. Occipital
   c. Sphenoid
   d. Ethmoid

36. The cardiovascular system regulates temperature by:
   a. Oxygenation and deoxygenation of the blood
   b. Contraction and relaxation of the heart
   c. Vasoconstriction and vasodilation of blood vessels
   d. All of the above

37. Macrophages:
   a. Are also phagocytes
   b. Can be 'wandering' or 'fixed'
   c. Develop from monocytes
   d. All of the above

38. Insulin cannot be given orally into the stomach because:
   a. The amount needed would cause the tablets to be too large
   b. The enzymes needed are found only in the skin
   c. It is a protein hormone and would be digested by enzymes there
   d. None of the above

39. A high fat, low fibre diet may predispose one to:
   a. Gallstones
   b. Cancer of the colon
   c. Appendicitis
   d. All of the above

40. An average heart rate would be between:
   a. 40 and 60 beats per minute
   b. 50 and 70 beats per minute
   c. 60 and 80 beats per minute
   d. 80 and 100 beats per minute

41. The four most abundant elements which make up about 96% of the body are:
   a. Carbon, oxygen, phosphorus, calcium
   b. Oxygen, nitrogen, sodium, carbon
   c. Carbon, oxygen, nitrogen, hydrogen
   d. Hydrogen, calcium, phosphorus, sodium

42. The process of oogenesis is:
   a. Production of male gametes
   b. Production of female gametes
   c. Production of male gonads
   d. Production of female gonads

43. The renal tubule includes the:
   a. Loop of Henle
   b. The distal convoluted tubule
   c. The proximal convoluted tubule
   d. All of the above

44. Yellow elastic tissue can be found in:
   a. Lung tissue
   b. Bronchial tubes
   c. Trachea
   d. All of the above

45. Which of the following cranial nerves is purely a sensory nerve?
   a. Optic (II)
   b. Facial (VII)
   c. Vagus (X)
   d. Hypoglossal (XII)

46. Ball and socket joints are found:
   a. At the knee and elbow only
   b. At the shoulder and hip only
   c. At the wrist and foot only
   d. None of the above

47. Centrioles are:
   a. Rod-like organelles that play an important part in mitosis
   b. Networks of fluid-filled cisterns for chemical reactions and transportation
   c. Vesicles containing digestive enzymes to break down and destroy foreign substances
   d. Chromosomes massed together which separate during cell division

48. Mechanical digestion includes:
   a. Peristalsis
   b. Segmentation
   c. Stomach churning
   d. All of the above

49. A client who reports a painless enlargement of lymph nodes under her arm should:
   a. Do more exercise as the lymphatic system may be blocked
   b. Book some massage sessions
   c. Consult her GP immediately
   d. Change her diet

50. Enuresis is common in:
   a. Childhood
   b. Teenagers
   c. Middle age
   d. Old age.

Mock exam paper 11

1. Hormones are produced by endocrine glands in response to:
   a. Hormonal stimulation
   b. Chemical changes in the blood
   c. Neural stimulation
   d. All of the above

2. The mammary artery can be found in the region of the:
   a. Breast
   b. Back
   c. Abdomen
   d. Heart

3. Which of the following is a true statement?
   a. White blood cells are larger than red ones but there are less of them
   b. Red blood cells carry oxygen and have a vital role in defence
   c. White blood cells contain haemoglobin which gives them their pale colour
   d. There are several different types of red blood cell but only one type of white blood cell

4. Apocrine glands excrete:
   a. Watery sweat
   b. Oil
   c. Fatty or milky sweat
   d. Serous fluid

5. Which part of the urinary system transports urine from the kidneys to the bladder?
   a. Pelvis
   b. Ureter
   c. Cortex
   d. Urethra

6. There are _____ bones in the axial skeleton and _____ in the appendicular skeleton.
   a. 60 and 206
   b. 80 and 126
   c. 126 and 80
   d. 206 and 60

7. Which of the following statements is true?
   a. Enzymes either speed up or slow down the rate of chemical reactions
   b. Enzymes are not affected by the acidic environment of the gut
   c. The digestion of food relies on the presence of enzymes
   d. All of the above

8. In the male reproductive system sperm mature in the:
   a. Penis
   b. Epididymis
   c. Prostate gland
   d. Vas deferens

9. Mucosa-associated lymphoid tissue (MALT) is found in the:
   a. Gastrointestinal tract
   b. Respiratory airways
   c. Urinary tract
   d. All of the above

10. The kidneys are not affected by:
    a. Diabetes mellitus
    b. Diabetes insipidus
    c. Addison's disease
    d. Graves' disease

11. A person suffering from a severe infection would have _____ in their blood.
    a. An increased number of red cells
    b. A decreased number of white cells
    c. An increased number of white cells
    d. A decreased number of platelets

12. The virus which causes chickenpox can reactivate in later life as:
    a. Rubella
    b. Herpes zoster
    c. Herpes simplex
    d. Rubeola

13. The study of the nervous system is:
    a. Histology
    b. Nephrology
    c. Craniopathy
    d. Neurology

14. The outermost layer of connective tissue which encircles the whole muscle is the:
    a. Endomysium
    b. Epimysium
    c. Perimysium
    d. Deep fascia

15. The main central shaft of a long bone is the:
    a. Diaphysis
    b. Epiphysis
    c. Endosteum
    d. Periosteum

16. The function of the large intestine is:
    a. Absorption of water
    b. Defecation
    c. Absorption of Vitamin K
    d. All of the above

17. The function of the nose is to _____ the incoming air.
    a. Moisten
    b. Warm
    c. Filter
    d. All of the above

18. Popliteal nodes are located:
    a. In the abdomen
    b. In front of the elbow
    c. Behind the knee
    d. In the neck

19. Which of the following is a false statement?
    a. Progesterone is only produced in the second half of the menstrual cycle
    b. The menstrual flow occurs as levels of oestrogen and progesterone decline
    c. The secretory phase occurs just before the menstrual period
    d. The corpus luteum is produced under the influence of FSH

20. The Islets of Langerhans are found:
    a. In the scrotum
    b. In the pancreas
    c. Either side of the thyroid
    d. Above the pituitary

21. Organic substances are ones which contain:
    a. Carbon
    b. Oxygen
    c. Nitrogen
    d. Hydrogen

22. The neurotransmitter which is released at the neuromuscular junction and triggers a muscle action potential is:
    a. Noradrenaline
    b. Dopamine
    c. Acetylcholine
    d. Adrenaline

23. In the nephron, the afferent arteriole divides into the:
    a. Glomerulus
    b. Efferent arteriole
    c. Collecting duct
    d. Bowman's capsule

24. The brain is responsible for:
    a. Controlling reflex actions
    b. Conducting impulses from sensory receptors
    c. Making and storing memories
    d. None of the above

25. Which of the following may be caused by cigarette smoking?
    a. Bronchitis
    b. Emphysema
    c. Lung cancer
    d. All of the above

26. Lymph nodes are enclosed in:
    a. Dense connective tissue
    b. Smooth muscle tissue
    c. Transitional epithelium
    d. Neuroglial tissue

27. The left atrium receives _____ blood from the _____.
    a. Oxygenated, lungs
    b. Deoxygenated, body
    c. Oxygenated, right atrium
    d. Deoxygenated, right ventricle

Mock exam paper 11 cont.

28. Facial massage would be contra-indicated in the presence of:
    a. Ephelides
    b. Lentigines
    c. Herpes simplex
    d. Albinism

29. Which of the following substances remains in the renal tubule?
    a. Amino acids
    b. Uric acid
    c. Glucose
    d. All of the above

30. An enlarged prostate gland:
    a. Causes difficulty in passing urine
    b. Can progress to being cancerous
    c. Can cause more frequent urination
    d. All of the above

31. The mineral content of the blood is affected by:
    a. Oestrogen
    b. Aldosterone
    c. Testosterone
    d. All of the above

32. The body is always striving to maintain a stable internal environment by the process of:
    a. Haemostasis
    b. Haemolysis
    c. Homeostasis
    d. Haemopoiesis

33. If the Achilles tendon ruptures/tears what action will not be possible:
    a. Waving
    b. Walking
    c. Breathing
    d. Eating

34. In this fracture the broken ends of the bone pierce the skin and infection may enter the wound:
    a. Compound
    b. Simple
    c. Comminuted
    d. Greenstick

35. The 'intrinsic factor' is necessary for the absorption of Vitamin _____ from the ileum.
    a. B1
    b. D
    c. K
    d. B12

36. Spina bifida is:
    a. A condition of old age
    b. Present at birth
    c. Develops at puberty
    d. Triggered by the menopause

37. For normal expiration to take place:
    a. Atmospheric pressure must be greater than the pressure in the lungs
    b. The pressure in the lungs must be greater than atmospheric pressure
    c. The pressure must be equal in both areas
    d. The pressure varies depending on the amount of activity being done

38. Calcium, potassium, iron and chlorine are all examples of:
    a. Fatty acids
    b. Vitamins
    c. Monosaccharides
    d. Minerals

39. The largest organ of the body is the:
    a. Liver
    b. Heart
    c. Skin
    d. Brain

40. Which of the following conditions is the result of a clotting disorder?
    a. Myocardial infarction
    b. Haemophilia
    c. Epistaxis
    d. Pernicious anaemia

41. Congenital absence of a nail is called:
    a. Anonychia
    b. Paronychia
    c. Leuconychia
    d. Koilonychia

42. Which of the following muscles moves the thumb?
    a. Flexor carpi ulnaris
    b. Flexor digitorum superficialis
    c. Flexor carpi radialis
    d. Flexor pollicis brevis

43. Stones found in the urinary system are called renal _____.
    a. Cysts
    b. Calculi
    c. Calyces
    d. Cusps

44. In relation to the vertebral column the cervical curve develops when:
    a. A baby is able to walk
    b. A baby is able to stand up
    c. A baby is able to sit up.
    d. A baby is able to hold its head up

45. Which of the following is true?
    a. Digestion occurs at the brush border.
    b. Digested proteins enter the blood stream via the lacteals
    c. Crypts of Lieberkuhn are situated in the duodenum
    d. Brunner's glands are situated in the stomach

46. Dysmenorrhea means:
    a. Painful menstruation
    b. Infertility
    c. Lack of menstruation
    d. Inflammation in the pelvis

47. The cauda equina is part of the:
    a. Autonomic Nervous System
    b. Brain
    c. Spinal cord
    d. Meninges

48. Oxygen is transported in the blood by:
    a. Plasma
    b. Red blood cells
    c. Platelets
    d. White blood cells

49. Which condition affects the normal development of leucocytes and is known to invade organs, such as the liver and spleen, causing infection, bleeding and anaemia?
    a. Lymphoma
    b. AIDS
    c. SLE
    d. Leukaemia

50. Which of the following is a false statement?
    a. The amount of urine passed daily is always the same
    b. The amount of urine passed daily is affected by how much exercise we do
    c. The amount of urine passed daily is affected by our general state of health
    d. The amount of urine passed daily is affected by hormones.

Mock exam paper 12

1. The name of the membrane which encloses the lungs is the:
   a. Pharynx
   b. Pericardium
   c. Peritoneum
   d. Pleura

2. Breaking down large droplets of fat into smaller ones is known as:
   a. Deamination
   b. Mastication
   c. Emulsification
   d. Segmentation

3. Which of the following is not found in the dermis?
   a. Cells producing melanin
   b. Blood supply
   c. Cells producing histamine
   d. Sweat glands

4. When does the thymus gland begin to atrophy?
   a. After puberty
   b. Before puberty
   c. During childhood
   d. During the menopause

5. A girl of 14 is much smaller than her peer group and has average sized parents. She could be lacking in _____ (hormone).
   a. ACTH
   b. GH
   c. LH
   d. ADH

6. Which of the following substances should not normally be present in urine?
   a. Water
   b. Glucose
   c. Urea
   d. Uric acid

7. The process of stopping bleeding is called:
   a. Homeopathy
   b. Homeostasis
   c. Haemostasis
   d. Haemopoiesis

8. Which myofilament gives skeletal muscle its dark banding?
   a. Titin
   b. Myosin
   c. Actin
   d. Troponin

9. The hormone which is responsible for the onset of labour is:
   a. Hormone replacement therapy
   b. Prolactin
   c. Luteinising hormone
   d. Oxytocin

10. The study of body tissues is called:
    a. Anatomy
    b. Histology
    c. Cytology
    d. Physiology

11. Efferent neurones conduct impulses:
    a. From CNS to muscles and glands
    b. From muscles and glands to CNS
    c. From PNS to CNS
    d. From sensory receptors to PNS

12. Which of the following is an incorrect statement?
    a. Mechanical stress is beneficial to the skeletal system
    b. Being bedridden causes bones to become weaker
    c. Ageing causes loss of calcium from the bones
    d. Exercise makes the bones more likely to fracture

13. Which of the following statements is false regarding the respiratory system?
    a. The pharynx serves as both an air and a food passageway
    b. The pharynx provides a passageway between the larynx and the bronchi
    c. The larynx contains the vocal cords
    d. The larynx is made of hyaline cartilage

14. The enzymes which digest proteins are known as:
    a. Peptides
    b. Amino acids
    c. Proteases
    d. Substrates

15. The flow of lymph is helped by:
    a. Exercise
    b. Massage
    c. Pressure from skeletal muscles
    d. All of the above

16. The right ventricle sends blood to the:
    a. Lungs
    b. Left ventricle
    c. Right atrium
    d. Body

17. Muscle fibres with a high resistance to fatigue, loaded with mitochondria and rich in myoglobin will not be of great use to a:
    a. 100 metre runner
    b. Marathon runner
    c. Long distance swimmer
    d. Long distance cyclist

18. The correct term for the shoulder blade is the:
    a. Sternum
    b. Scapula
    c. Sesamoid
    d. Sacrum

19. Hyposecretion of the antidiuretic hormone causes:
    a. Addison's disease
    b. Cushing's syndrome
    c. Diabetes insipidus
    d. Diabetes mellitus

20. Synaptic vesicles contain:
    a. Myelin
    b. Neurotransmitters
    c. Schwann cells
    d. Neurilemma

21. Androgens are produced by:
    a. Testes, ovaries, placenta
    b. Ovaries, adrenal glands in women, adrenal glands in men
    c. Adrenal glands in men, adrenal glands in women, testes
    d. Testes, placenta, adrenal glands in women

22. The axillary lymph nodes are situated in the:
    a. Armpits
    b. Groin
    c. Abdomen
    d. Neck

23. Which of the following can be caused by infection and inflammation of the alveoli and surrounding tissues?
    a. Pleurisy
    b. Pneumonia
    c. Hay fever
    d. SARS

24. The function of hydrochloric acid is:
    a. To activate pepsin
    b. An antibacterial
    c. To inhibit salivary amylase
    d. All of the above

25. What happens to the hair during telogen?
    a. Growing
    b. Shedding
    c. Changing
    d. Dividing

26. The afferent arteriole bringing blood to the glomerulus is wider than the efferent arteriole because:
    a. It is nearer to the renal artery
    b. It causes high pressure within the glomerulus
    c. It is carrying many substances to be filtered
    d. It causes the large plasma proteins to be filtered out of the blood

27. The insertion part of a muscle is usually:
    a. Attached to the bone which moves
    b. The fixed point
    c. The origin of the muscle
    d. Attached to the stationary bone

Mock exam paper 12 cont.

28. The way in which bone tissue is continually replaced throughout life is called:
    a. Retraction
    b. Replacement
    c. Reincarnation
    d. Remodelling

29. Which of the following is true?
    a. Exercise increases the rate and depth of breathing because more carbon dioxide is produced
    b. Exercise increases the rate and depth of breathing because the brain sends messages to the diaphragm to increase its activity
    c. Exercise increases the rate and depth of breathing because there are receptors which monitor the amount of oxygen and carbon dioxide in the blood
    d. All of the above

30. Myoedema is not caused by:
    a. Hypothyroidism
    b. Surgical removal of too much thyroid tissue
    c. Hyperthyroidism
    d. Too little thyroxine

31. We can combat the effect of free radicals by consuming more foods containing:
    a. Fats
    b. Water
    c. Antibiotics
    d. Antioxidants

32. Which of the following statements regarding the liver is false?
    a. It is made up of 3 lobes arranged in lobules
    b. It is made up of hepatocytes
    c. It contains spaces called sinusoids
    d. It contains phagocytes

33. A viral infection which lowers an individual's immunity and makes them more susceptible to infections is:
    a. Systemic lupus erythematosus
    b. Infectious mononucleosis
    c. Human immunodeficiency virus
    d. Viral encephalitis

34. Which of the following blood vessels carries oxygenated blood?
    a. Pulmonary artery
    b. Pulmonary vein
    c. Coronary sinus
    d. None of the above

35. Which fact is incorrect regarding the male urethra?
    a. Is the terminal duct of the male urinary system
    b. Transports semen and urine out of the body
    c. Is the terminal duct of the male reproductive system
    d. Is a 4 cm long tube lying behind the symphysis pubis

36. Bell's palsy is weakness or paralysis of the _____?
    a. Arms
    b. Legs
    c. Face
    d. Bladder

37. The cause of rheumatoid arthritis is thought to be:
    a. Due to injury
    b. Autoimmune
    c. Bacterial
    d. Age

38. The dermal papillae:
    a. Increase the surface area of the papillary layer
    b. Are responsible for fingerprints
    c. Contain capillary networks
    d. All of the above

39. The odd one out below is:
    a. Lunula
    b. Acid mantle
    c. Matrix
    d. Cuticle

40. An autoimmune disease characterised by weakness of skeletal muscles, especially drooping eyelids, could be:
    a. Myasthenia gravis
    b. Muscular rheumatism
    c. Fibrositis
    d. Metastasis

41. Which of the following is the correct term for 'water-loving'?
    a. Hydrophobic
    b. Carbohydrate
    c. Hydrophilic
    d. Dehydration

42. White spots or streaks on nails is called:
    a. Paronychia
    b. Anonychia
    c. Leuconychia
    d. Koilonychia

43. Antibodies being passed on in breast milk from mother to baby is an example of:
    a. Naturally acquired passive immunity
    b. Artificially acquired active immunity
    c. Naturally acquired active immunity
    d. Artificially acquired passive immunity

44. Adrenaline and noradrenaline are released by the:
    a. Mid brain
    b. Adrenal cortex
    c. Medulla oblongata
    d. Adrenal medulla

45. Which of the following statements regarding osmoreceptors is false?
    a. They detect a decreased level of water in the blood
    b. They detect an increased amount of solutes in the blood
    c. They are situated in the hypothalamus in the brain
    d. They are situated in the medulla oblongata in the brain

46. Which is a false statement regarding the sense of smell?
    a. It is interpreted in the brain by the facial nerve
    b. It is closely linked with the sense of taste
    c. It can be affected by a frontal head injury
    d. It is closely linked to emotions and memories

47. The saphenous veins are to be found in the:
    a. Trunk
    b. Arms
    c. Legs
    d. Head

48. Inability to conceive may be caused by:
    a. Repeated infections of the Fallopian tubes
    b. Impotence
    c. Endometriosis
    d. All of the above

49. Which of the following is a false statement?
    a. Uraemia causes drowsiness, lethargy and nausea
    b. Uraemia is the presence of blood in the urine
    c. Uraemia can be fatal if left untreated
    d. Uraemia can be caused by renal failure

50. Cardiac muscle cells are described as being 'myogenic'. This means:
    a. They have a constant rhythm
    b. The impulse to beat comes from the endocrine system
    c. They can contract independently of the nervous system
    d. Each cell contracts independently at different times.

Mock exam paper 13

1. The liver does not:
   a. Store vitamin C
   b. Store iron and copper
   c. Convert substances into glucose
   d. Convert ammonia into urea
2. Breathing cannot be controlled during:
   a. Singing
   b. Sleeping
   c. Swimming
   d. Shouting
3. During muscle contraction the:
   a. Sarcomeres lengthen
   b. Myofilaments shorten
   c. Sarcomeres shorten
   d. Myofilaments lengthen
4. Killer cells:
   a. Kill tumour cells in the body
   b. Kill the body's own contaminated cells
   c. Kill invading cells
   d. All of the above
5. The testes lie outside the body because:
   a. The internal secretions of the body are too acidic
   b. They are under the influence of the Y chromosome
   c. Sperm survive better at lower temperatures
   d. All of the above
6. Impulses or messages are conducted _____ within the nervous system.
   a. Electrically only
   b. Chemically only
   c. Both electrically and chemically
   d. Neither electrically nor chemically
7. Blood pressure is not affected by the actions of:
   a. Renin
   b. Vasopressin
   c. Sodium
   d. Calcitriol
8. Carbohydrates do not contain:
   a. Carbon
   b. Oxygen
   c. Nitrogen
   d. Hydrogen
9. Yellow bone marrow is found in the:
   a. Epiphysis
   b. Diaphysis
   c. Periosteum
   d. None of the above

10. Which of these is a false statement?
    a. Women do not produce androgens
    b. Cortisol helps to deal with long term stress
    c. Men do not produce progesterone
    d. Adrenaline is known as the 'fight or flight' hormone
11. Which of the following statements is false?
    a. The heart valves are composed of connective tissue covered by endocardium
    b. The atrioventricular valve on the left has 2 cusps but 3 on the right
    c. The atrioventricular valves are called the aortic and pulmonary valves
    d. The semilunar valves lie between the ventricles and the 2 main arteries leaving the heart
12. Adipocytes are found in the:
    a. Reticular dermis
    b. Subcutaneous layer
    c. Papillary dermis
    d. Epidermis
13. The pyloric sphincter lies between the:
    a. Oesophagus and the stomach
    b. Rectum and the anus
    c. Stomach and the duodenum
    d. Bile duct and the duodenum
14. Most muscle fibres in the body are composed of:
    a. Type I (slow oxidative)
    b. Type II (fast oxidative)
    c. Type II (fast glycolytic)
    d. A mixture of all types
15. The function of relaxin is to:
    a. Increase flexibility of the ligaments and joints during labour
    b. Control the menstrual cycle
    c. Control the process of spermatogenesis
    d. Decrease the risk of premature labour
16. Which is a false statement regarding the blood-brain barrier?
    a. Main purpose is to protect the brain cells
    b. Allows alcohol to pass directly to the brain cells
    c. Does not allow some beneficial drugs to pass to the brain cells
    d. Protects the brain cells from nicotine and caffeine
17. All cells require ATP. ATP is the _____ of the cell.
    a. Protein manufacturer
    b. Transport centre
    c. Energy currency
    d. Hereditary units

18. Which of the following are irregular bones?
    a. Facial and vertebrae
    b. Upper and lower limbs
    c. Skull and pelvis
    d. Wrists and ankles
19. The way the lower respiratory system is organised resembles, and is often referred to as, the bronchial _____.
    a. Root
    b. Leaf
    c. Tree
    d. Trunk
20. The causes of lymphoedema can be:
    a. Congenital abnormality of lymphatic vessels
    b. Blockage to lymphatic vessels by a parasite
    c. Damage to lymphatic vessels following surgery
    d. All of the above
21. Which of the following statements regarding the urethra is true?
    a. It is shorter in women
    b. It transports both urine and semen in men
    c. It connects the bladder to the outside
    d. All of the above are true
22. Type I diabetes:
    a. Is not usually treated with insulin
    b. Is a disease of old age
    c. Is thought to be an autoimmune disease
    d. Can be treated with a low protein diet
23. The sound created by the beating heart is due to the:
    a. Contraction of the ventricles
    b. Closing of the valves
    c. Movement of blood
    d. Contraction of the atria
24. Which of the following is not a skin infestation?
    a. Pediculosis corporis
    b. Tinea corporis
    c. Scabies
    d. Pediculosis capitis
25. The medulla oblongata:
    a. Regulates water balance and temperature balance
    b. Controls movement, posture and balance
    c. Controls cardiac and respiratory centres
    d. Receives and processes sensory information

Mock exam paper 13 cont.

26. When an enzyme is _____ it can no longer function.
    a. Denatured
    b. Deaminated
    c. Detoxified
    d. Defecate

27. An example of an autoimmune disease is:
    a. Diabetes mellitus
    b. Rheumatoid arthritis
    c. Multiple sclerosis
    d. All of the above

28. External respiration occurs:
    a. In the lungs
    b. Between lungs and blood
    c. In cells
    d. Between blood and tissue cells

29. During walking, if the quadriceps are the prime movers then the _____ will be the antagonists.
    a. Gluteals
    b. Adductors
    c. Hamstrings
    d. Obliques

30. The secondary sex characteristics develop:
    a. At puberty in both boys and girls
    b. At the menarche in both boys and girls
    c. At the age of 11 in both boys and girls
    d. At the age of 14 in both boys and girls

31. The three bones which have fused together to form the innominate bone are:
    a. Ischium, pubis, ilium
    b. Ilium, sacrum, ischium
    c. Pelvis, sacrum, coccyx
    d. Pubis, sacrum, ilium

32. The adrenal glands are found:
    a. In the throat
    b. Above the kidneys
    c. In the abdomen
    d. Below the brain

33. The hepatic portal circulation involves the:
    a. Heart and the rest of the body
    b. Heart and the brain
    c. Heart and the lungs
    d. Heart and the digestive system

34. Starch, disaccharides and lactose are all examples of:
    a. Proteins
    b. Enzymes
    c. Carbohydrates
    d. Lipids

35. A reflex action:
    a. Is under the control of the endocrine system
    b. Is under the control of the will
    c. Occurs without the involvement of the spinal cord
    d. Occurs without the involvement of the brain

36. Miliaria rubra (prickly heat) is a disorder of the:
    a. Sebaceous glands
    b. Nail bed
    c. Sweat glands
    d. Hair follicle

37. The cause of cystic fibrosis is:
    a. Genetic
    b. Viral
    c. Cancer
    d. Age

38. The soleus muscle lies _____ to the gastrocnemius muscle.
    a. Medial
    b. Deep
    c. Superficial
    d. Proximal

39. Which of these groups of lymph nodes are not found in the head and neck?
    a. Submandibular
    b. Axillary
    c. Occipital
    d. Parotid

40. The correct order of events in somatic cell division is:
    a. Metaphase, anaphase, telophase, prophase
    b. Telophase, prophase, anaphase, metaphase
    c. Prophase, metaphase, anaphase, telophase
    d. Anaphase, metaphase, telophase, prophase

41. Which of the following would indicate a 'normal' range of blood pressure in a healthy person?
    a. Between 70 and 100 systolic and 20 and 50 diastolic
    b. Between 100 and 130 systolic and 60 and 90 diastolic
    c. Between 130 and 160 systolic and 90 and 120 diastolic
    d. Between 160 and 190 systolic and 120 and 150 diastolic

42. Hair loss can be caused by:
    a. Hormonal change
    b. Poor nutrition
    c. Emotional stress
    d. All of the above

43. In a cell-mediated response:
    a. Cells attack cells
    b. Cells attack antibodies
    c. Antibodies attack cells
    d. All of the above

44. Medication which can help women to cope with the effects of the menopause is:
    a. PID
    b. STD
    c. HRT
    d. PMT

45. The colour of urine is determined by:
    a. Bile
    b. Blood
    c. Hormones
    d. Toxins

46. Pronation and supination refer to movements of the:
    a. Hands
    b. Knees
    c. Neck
    d. Abdomen

47. Hypersecretion from the adrenal cortex will not cause:
    a. Virilism
    b. Cushing's syndrome
    c. Diabetes insipidus
    d. Hirsutism

48. This part of the nail secures the nail to the finger or toe:
    a. Nail groove
    b. Nail plate
    c. Nail wall
    d. Nail bed

49. Each ureter drains the renal pelvis of a kidney and inserts into the _____ aspect of the bladder.
    a. Anterior
    b. Posterior
    c. Superior
    d. Inferior

50. Which of the following statements is false?
    a. Cutaneous membranes are found in joints
    b. Serous membranes do not open directly to the exterior
    c. Synovial membranes secrete synovial fluid
    d. A function of mucous membranes is protection.

Mock exam paper 14

1. **Which of the following glands produce oil to protect the skin and hair?**
   a. Sebaceous
   b. Sweat
   c. Apocrine
   d. Eccrine

2. **Light rays are received in the eye by photoreceptors on the _____.**
   a. Aqueous humour
   b. Retina
   c. Vitreous humour
   d. Lens

3. **Which of the following is not a function of the vertebral column?**
   a. Support the head
   b. Protect the spinal cord
   c. Allow attachment for the ribs
   d. Protects the heart and lungs

4. **Infants are tested for adequate thyroid function at birth in order to prevent:**
   a. Graves' disease
   b. Cretinism
   c. Goitre
   d. Cushing's disease

5. **Which of the following is false regarding the pancreas?**
   a. It is both an endocrine and an exocrine gland
   b. Many different enzymes are made and secreted by the pancreas
   c. Pancreatic juice is secreted from about a third of the pancreatic cells
   d. Pancreatic juice is of an alkaline nature

6. **The function of the 'stirrup' is to:**
   a. Connect the middle ear with the upper portion of the throat
   b. Carry sound waves from the auricle to the eardrum
   c. Transmit sound waves to the inner ear
   d. Secrete cerumen

7. **The inguinal lymph nodes are situated in the:**
   a. Armpits
   b. Groin
   c. Abdomen
   d. Neck

8. **Immunocompetent cells are:**
   a. B cells
   b. T cells
   c. Both B and T cells
   d. Neither B and T cells

9. **Most of the respiratory passages are lined with:**
   a. Ciliated epithelium
   b. Columnar epithelium
   c. Cuboid epithelium
   d. Cartilage

10. **Which layer of the heart wall contracts to pump blood?**
    a. Endocardium
    b. Pericardium
    c. Myocardium
    d. Epicardium

11. **Reproductive cells are called:**
    a. Gametes
    b. Meiosis
    c. Gonads
    d. Zygote

12. **The Organ of Corti is found in the:**
    a. Eye
    b. Tongue
    c. Ear
    d. Nose

13. **An example of a condyloid joint can be found at the:**
    a. Wrist
    b. Ankle
    c. Elbow
    d. Shoulder

14. **Metastasis means:**
    a. An allergic reaction to a metal
    b. The spread of cancer
    c. An infection of a lymph node
    d. The inability to destroy cancer cells

15. **The iliopsoas muscle:**
    a. Abducts the hip joint
    b. Extends the hip joint
    c. Adducts the hip joint
    d. Flexes the hip joint

16. **Rennin is an enzyme found only in the stomachs of:**
    a. Teenagers
    b. Women
    c. Infants
    d. Elderly people

17. **Addison's disease is caused by:**
    a. Hypersecretion from the thyroid gland
    b. Hypersecretion from the ovaries
    c. Hyposecretion from the adrenal cortex
    d. Hyposecretion from the parathyroids

18. **Electrolytes in the blood are:**
    a. Balanced by the kidneys
    b. Substances such as sodium, chloride or potassium
    c. Affected by the water balance in the body
    d. All of the above

19. **Macrophages within the alveoli are present in order to:**
    a. Prevent infection
    b. Maximise oxygen uptake
    c. Prevent collapse
    d. Keep them moisturised

20. **Which of the following is a false statement?**
    a. Veins have a tripled-layer wall just like arteries
    b. Veins have valves in their walls
    c. The lumen of a vein is usually smaller than an artery
    d. Veins usually carry blood towards the heart

21. **This is a common bacterial infection, very contagious and characterised by weeping yellow blisters or crusty sores:**
    a. Impetigo
    b. Furuncle
    c. Hordeolum
    d. Boil

22. **Vaccination is giving a _____ form of a disease to a person:**
    a. High energy
    b. Artificial
    c. Weakened
    d. Inactive

23. **Eccrine sweat glands do not produce:**
    a. Urea
    b. Salts
    c. Hormones
    d. Water

24. **The sensory receptors in the olfactory epithelium are:**
    a. Gustatory receptors
    b. Photoreceptors
    c. Mechanoreceptors
    d. Chemoreceptors

25. **Deoxygenated blood is brought to the lungs for gaseous exchange by the:**
    a. Pulmonary veins
    b. Pulmonary arteries
    c. Bronchial arteries
    d. Bronchial veins

26. **_____ is essential for muscle contraction.**
    a. Carbon
    b. Nitrogen
    c. Calcium
    d. Hydrogen

27. **Lipids arrive at the liver via:**
    a. The hepatic artery
    b. The hepatic portal vein
    c. Both of the above
    d. Neither of the above

Mock exam paper 14 cont.

28. Movements at the feet can be:
    a. Inversion
    b. Dorsiflexion
    c. Plantarflexion
    d. All of the above

29. Which of the following is true?
    a. Passive immunity is when the body makes its own antibodies
    b. Vaccinations achieve immunity without recipients being ill
    c. Immunity lasts longer when ready made antibodies are used
    d. Naturally acquired active immunity involves vaccination

30. 'Blood pressure' is:
    a. The average amount of blood contained in the heart at any one time
    b. The force with which the valves are able to close
    c. The combined pressure of blood in the venous and capillary system
    d. The force exerted by the blood upon the walls of a blood vessel

31. An autoimmune disease affecting the myelin sheaths of nerves with periods of remission is likely to be:
    a. Myalgic encephalomyelitis
    b. Motor neurone disease
    c. Meningitis
    d. Multiple sclerosis

32. The innermost layer of the pleura is the _____ layer.
    a. Parietal
    b. Peripheral
    c. Visceral
    d. Varicose

33. Which of the following muscles is a muscle of facial expression?
    a. Buccinator
    b. Risorius
    c. Temporalis
    d. Pterygoids

34. The temporo-mandibular joint is a _____ joint.
    a. Diarthrotic
    b. Synarthrotic
    c. Amphiarthrotic
    d. None of the above

35. The hypothalamus:
    a. Regulates posture and balance
    b. Coordinates sneezing and coughing
    c. Controls sensations of hunger and thirst
    d. Helps to control respiration

36. Nocturia may occur:
    a. At night
    b. In pregnant women
    c. Due to enlargement of the prostate gland
    d. All the above

37. Goblet cells in the digestive system secrete:
    a. Digestive juices
    b. Mucous
    c. Enzymes
    d. Saliva

38. The term 'superficial' means:
    a. At the front of the body
    b. At the back of the body
    c. Towards the surface of the body
    d. Away from the surface of the body.

39. At the capillary bed:
    a. Solutes from the blood pass into the interstitial fluid
    b. Pressure at the arteriolar end is higher than at the venous end
    c. Waste products can pass through the capillary wall because it is only one cell thick
    d. All of the above

40. A spasm could be defined as:
    a. An abnormal, involuntary muscular contraction
    b. A sudden painful contraction of a muscle
    c. Overstretching of a muscle
    d. Tearing of muscle fascia

41. Which of the following hormones is released by the Anterior Pituitary Gland?
    a. Antidiuretic hormone
    b. Thyroid-stimulating hormone
    c. Glucocorticoids
    d. All of the above

42. Which is the correct order of events regarding inspiration?
    a. Air is drawn into the lungs, the diaphragm contracts, the pressure in the thorax decreases, the volume in the thorax increases
    b. The volume in the thorax increases, air is drawn into the lungs, the pressure in the thorax decreases, the diaphragm contracts
    c. The diaphragm contracts, the volume in the thorax increases, the pressure in the thorax decreases, air is drawn into the lungs
    d. The pressure in the thorax decreases, the diaphragm contracts, air is drawn into the lungs, the volume in the thorax increases

43. A severe cramping pain in the lower back could be:
    a. Incontinence
    b. Renal colic
    c. Anuria
    d. Pyelitis

44. Cystitis commonly affects women because:
    a. Women have shorter ureters
    b. Women have weaker bladders
    c. Women have shorter urethras
    d. Women have smaller kidneys

45. The hormone responsible for regulating the mineral content of the blood is:
    a. Glucagon
    b. Mineralcorticoids
    c. Glucocorticoids
    d. Thymosin

46. What is the function of areolar tissue?
    a. To surround body organs
    b. To move substances along a passageway
    c. To line blood and lymphatic vessels
    d. To allow organs to distend or expand

47. Which of the following protects the nail bed from infection?
    a. Cuticle
    b. Eponychium
    c. Hyponychium
    d. Nail wall

48. An imaginary line that divides the body vertically into right and left portions is called the:
    a. Sagittal plane
    b. Frontal plane
    c. Oblique plane
    d. Transverse plane

49. The function of the vas deferens is to:
    a. Store sperm until they are fully mature
    b. Transport sperm from the epididymis to the urethra
    c. Transport semen to the exterior of the body
    d. Produce sperm through the process of spermatogenesis

50. A basic structured unit called a Haversian system is found in:
    a. The aveoli
    b. A neurone
    c. Myofibrils
    d. Compact bone tissue.

Answer key

Please note that no answers are provided for the revision and self-study sections that form the first part of each chapter.

Chapter 1
Page 11 Exercise 1
1. Lateral
2. Medial
3. Superior, cranial or cephalad
4. Proximal
5. Distal
6. Superficial
7. Deep
8. Inferior or caudal
9. Midline or median line

Page 12 Exercise 2
a. Superior
b. Inferior
c. Medial
d. Lateral
e. Proximal
f. Distal
g. Periphery

Exercise 3
1. h
2. d
3. b
4. g
5. c
6. j
7. a
8. e
9. i
10. f

Exercise 4
Dorsal cavity – Brain, spinal cord
Thoracic cavity – Oesophagus, heart, lungs
Abdominal cavity – Gall bladder, liver, pancreas
Pelvic cavity – Urinary bladder, reproductive organs

Page 13 Exercise 5
a. Sagittal plane
b. Frontal/coronal plane
c. Transverse plane
d. Oblique plane

Vocabulary test
a. The study of the structure of the body
b. Towards the head, above
c. Divides horizontally into inferior and superior portions
d. Farther from its origin or point of attachment of a limb
e. Line through the middle of the body
f. Relating to the inner walls of a body cavity
g. The study of the diseases of the body
h. The study of the functions of the body
i. At the front of the body, in front of
j. Relating to the internal organs of the body

Page 14 Multiple choice questions
1. c
2. a
3. b
4. d
5. b
6. d
7. a
8. c
9. a
10. b

Chapter 2
Page 20 Exercise 1
Across
Magnesium
Calcium
Phosphorous
Nitrogen
Iodine
Chlorine

Down
Hydrogen
Sodium
Iron
Oxygen
Carbon
Sulphur
Potassium

Exercise 2
a. Inorganic, water, temperature, lubricant, brain, solvent

b. carbon, lipids, proteins, nucleic acids, adenosine triphosphate
c. carbon, hydrogen, fuel/energy
d. fats, carbon, hydrogen, oxygen
e. carbon, hydrogen, oxygen, nitrogen, sulphur

Page 21 Exercise 3
1. f
2. h
3. k
4. c
5. e
6. g
7. j
8. d
9. a
10. b
11. i

Page 22 Exercise 4
a. Lysosome
b. Smooth endoplasmic reticulum
c. Nucleus
d. Site of protein synthesis
e. Barrier that surrounds the cell
f. Golgi body
g. Cytoplasm

Exercise 5
a. T
b. F
c. F
d. T
e. F
f. F
g. T
h. F
i. F
j. T

Page 23 Exercise 6
a. Areolar tissue
b. Layers of cells that change shape to allow tissue to stretch
c. Intervertebral discs
d. Secretion and absorption
e. Ciliated simple columnar epithelium
f. Insulation, energy reserve, support and protection
g. Elastic connective tissue/yellow elastic tissue
h. Heart wall
i. Mucous membranes
j. Lubrication and cushioning

Page 24 Vocabulary test
a. The engulfment and destruction of microbes, cell debris and foreign matter by phagocytes
b. A protein that speeds up the rate of a reaction without itself being used in the reaction
c. The process by which the body maintains a stable internal environment
d. Having two complete sets of chromosomes per cell
e. A small, fluid-filled sac
f. The changes that take place within the body to enable its growth and function
g. Any cell except the reproductive cells
h. The basic unit of genetic material
i. A liquid in which a solid is dissolved
j. The energy of motion

Page 25
Multiple choice questions
1. c
2. d
3. b
4. a
5. c
6. a
7. d
8. a
9. b
10. c
11. a
12. a
13. b
14. a
15. c

Chapter 3
Page 36 Exercise 1
NB: 10 functions have been listed here but you only need to give 7 of them
a. Heat regulation
b. Sensation
c. Protection
d. Absorption
e. Excretion
f. Secretion
g. Vitamin D synthesis
h. Immunity
i. Blood reservoir
j. Communication

Exercise 2
a. Hair shaft
b. Sebaceous gland
c. Arrector pili muscle
d. Hair root
e. Hair follicle
f. Epidermis
g. Dermis
h. Subcutaneous tissue

Exercise 3
a. F
b. T
c. F
d. T
e. F

Page 37 Exercise 4
1. d
2. f
3. b
4. g
5. c
6. h
7. a
8. e

Exercise 5
The correct order is 4, 2, 5, 1, 3

Exercise 6
a. dermis, papillary, reticular
b. papillary, areolar connective, elastic, papillae, capillaries
c. reticular, dense irregular connective, collagen, elastic, hair, nerves, glands, sweat, adipose, reticular, subcutaneous, areolar connective, adipose

Page 38 Exercise 7
a. dry
b. normal/balanced
c. sensitive

Exercise 8
Across
3. Protection
7. Root
9. Lanugo
12. Catagen
13. Eponychium
16. Telogen

Down
1. Fold
2. Oil
3. Pili
4. Nostril
5. Shaft
6. Lunula
8. Terminal
10. Anagen
11. Papilla
14. Onyx
15. Bed

Page 39 Exercise 9
a. 2
b. 3
c. 2
d. 1
e. 1

Page 40 Exercise 10
a. Eczema
b. Furuncle (boil)
c. Tinea capitis (scalp ringworm)
d. Tinea pedis (athlete's foot)
e. Chloasma
f. Vitiligo
g. Milia
h. Warts
i. Beau's lines
j. Paronychia
k. Tinea ungium (ringworm of the nail)

Page 41 Exercise 11
1. c
2. f
3. h
4. n
5. a
6. p
7. l
8. q
9. o
10. k
11. e
12. m
13. g
14. i
15. d
16. j
17. b

Vocabulary test
a. Cross-infection
b. Desquamation
c. Integumentary system
d. Vasoconstriction
e. Vasodilation

Page 42
Multiple choice questions
1. c
2. d
3. b
4. b
5. a
6. d
7. c
8. c
9. d
10. c
11. a
12. c
13. a
14. b
15. c

Chapter 4
Page 57 Exercise 1
Across
Energy storage
Haemopoiesis
Movement

Down
Mineral homeostasis
Support
Protection

Exercise 2
a. Osteology, osseous, connective
b. mineral, collagen, calcium, hydroxyapatite, hardness

Exercise 3
a. Compact (dense) bone tissue
b. Spongy (cancellous) bone tissue

Page 58 Exercise 4
a. Long bone
b. Short bone
c. Flat bone
d. Irregular bone
e. Sesamoid bone

Exercise 5
1. d
2. f

3. b
4. g
5. c
6. a
7. e

Page 59 Exercise 6
a. axial, appendicular

Exercise 7
a. Frontal bone
b. Nasal bone
c. Maxilla
d. Zygomatic bone
e. Temporal bone
f. Inferior nasal concha (turbinate)
g. Mandible
h. Lacrimal bone
i. Vomer
j. Ethmoidal bone
k. Parietal bone
l. Sphenoidal bone

Page 60 Exercise 7 cont
a. Frontal bone
b. Parietal bone
c. Occipital bone
d. Nasal bone
e. Lacrimal bone
f. Zygomatic bone
g. Sphenoidal bone
h. Ethmoidal bone
i. Temporal bone
j. Maxilla
k. Mandible

Exercise 8
a. Seven
b. Twelve
c. Five
d. Five
e. Four

Page 61 Exercise 9
a. Hyoid
b. Sternum
c. Ribs

Exercise 10
a. Clavicle
b. Scapula
c. Humerus
d. Radius
e. Ulna

Page 62 Exercise 11
a. Ilium
b. Femur
c. Patella
d. Fibula
e. Tibia

Page 63 Exercise 12
Tarsals
Talus
Medial cuneiform
Calcaneus
Navicular
Cuboid
Lateral cuneiform
Intermediate cuneiform

Carpals
Scaphoid
Triquetrum
Pisiform
Hamate
Trapezoid
Lunate
Capitate
Trapezium

Exercise 13
a. F
b. T
c. F
d. T
e. T
f. F

Exercise 14
a. Flexing
b. Abducting
c. Supinating
d. Circumduct

Page 64 Exercise 15
a. Condyloid (ellipsoid) joint
b. Hinge joint
c. Saddle joint
d. Pivot joint
e. Gliding (plane) joint
f. Ball and socket (spheroidal) joint

Exercise 16
a. Prolapsed intervertebral disc
b. Rheumatoid arthritis
c. Greenstick
d. Gout
e. Kyphosis
f. Osteoporosis

g. Lordosis
h. Hammer toes
i. Pes cavus
j. Scoliosis

Page 65 Vocabulary test
a. A tough band of connective tissue that attaches muscles to bones
b. The process of bone formation
c. A tough band of connective tissue that attaches bones to bones
d. The production of blood cells and platelets
e. The point of contact between two bones – a joint

Multiple choice questions
1. b
2. a
3. d
4. a
5. c
6. c
7. a
8. d
9. c
10. a
11. d
12. b
13. b
14. d
15. d

Chapter 5
Page 74 Exercise 1
Across
3. ATP
4. Sarcomere
7. Epimysium
10. Glucose
11. Endomysium
14. Locomotion
16. Thermogenesis
17. Actin
18. Oxygen

Down
1. Calcium
2. Tone
5. Fatigue
6. Visceral
8. Sarcolemma
9. Myofibre
12. Elasticity

13. Cardiac
15. Atony

Page 75 Exercise 2
a. F
b. T
c. T
d. F
e. T
f. F
g. T
h. T
i. T
j. F

Exercise 3
a. isotonic, isometric, isotonic, isometric
b. Isotonic, isometric
c. oxygen

Page 76 Exercise 4
a. Smallest
b. Walking and running
c. White

Exercise 5
a. 4
b. 3
c. 1
d. 5
e. 2

Exercise 6
a. Occipitalis
b. Orbicularis oculi
c. Orbicularis oris
d. Buccinator
e. Mentalis
f. Sternocleidomastoid

Page 77 Exercise 7
Pectoralis minor
Serratus anterior
Rectus abdominis
Pectoralis major

Exercise 8
Back
Trapezius
Erector Spinae
Latissimus dorsi

Arm
Deltoid
Brachialis

Biceps brachii
Pronator teres
Supinator

Leg
Gracilis
Sartorius
Soleus
Gastrocnemius
Tibialis anterior
Biceps femoris

Exercise 9
1. e
2. f
3. g
4. b
5. d
6. c
7. a

Page 78 Exercise 10
a. T
b. F
c. T
d. F
e. F
f. F

Vocabulary test
a. Aponeurosis
b. Atrophy
c. Fascia
d. Fibrosis
e. Hypertonia
f. Insertion
g. Origin
h. Tendon

Multiple choice questions
1. b
2. b
3. a
4. d
5. d
6. a
7. c
8. d
9. b
10. d
11. c
12. a
13. d
14. c
15. b

Chapter 6
Page 90 Exercise 1
Central nervous system (CNS)
The brain and spinal cord belong to the CNS. Its function is processing and integrating information

Peripheral nervous system (PNS)
The cranial and spinal nerves belong to the PNS. Its function is carrying impulses

Page 91 Exercise 2
a. Somatic
b. Skeletal
c. Autonomic
d. Inhibits
e. Stimulates

Exercise 3
a. T
b. F
c. F
d. T
e. T
f. F

Page 92 Exercise 4
a. Cell body
b. The receiving or input portion of a neurone
c. The transmitting portion of a neurone
d. Neurotransmitter
e. Synaptic vesicle
f. Sheath that protects and insulates neurones and speeds up the conduction of nerve impulses
g. Schwann cells (neurolemmocytes)
h. Gaps in the myelin sheath

Exercise 5
a. Diencephalon
b. Brain stem – Midbrain, Pons, Medulla oblongata
c. Cerebrum
d. Cerebellum
e. Spinal cord

Page 93 Exercise 6
a. Dura mater
b. Ventricles
c. Meninges
d. Arachnoid

Exercise 7 Across
Limbic system
Cerebrum
Brain stem
Vagus
Abducens
Olfactory

Down
Cerebellum
Diencephalon
Twelve
Cerebral cortex

Page 94 Exercise 8
1. b
2. d
3. e
4. a
5. c

Exercise 9
a. 2
b. 4
c. 1
d. 3

Page 95 Exercise 10
a. outer/external, auricle/pinna, tympanic, vibrates, middle
b. middle, eardrum/tympanic membrane, oval, round, hammer/malleus, anvil/incus, stirrup/stapes, auditory ossicles, Eustachian/auditory tube
c. inner, perilymph, endolymph, vestibule, canals, Corti

Exercise 11
a. T
b. F
c. T
d. T
e. F

Page 96 Exercise 12
1. Meningitis
2. Epilepsy
3. Anosmia
4. Alzheimer's disease
5. Spina bifida
6. Bell's palsy
7. Neuritis
8. Sciatica
9. Concussion
10. Glaucoma

Page 97 Vocabulary test
a. Ganglion
b. Meninges
c. Neurology
d. Proprioceptor

Multiple choice questions
1. c
2. b
3. a
4. b
5. a
6. d
7. c
8. a
9. d
10. c
11. a
12. c
13. c
14. b
15. a

Chapter 7
Page 102 Exercise 1
Across
1. Pineal
3. Parathyroid
6. Hypothalamus
7. Exocrine
10. Testes
11. Hormone
13. Adrenal
14. Thyroid

Down
1. Pituitary
2. Pancreas
4. Ovary
5. Endocrine
8. Reproduction
9. Medulla
12. Nervous
14. Thymus

Page 103 Exercise 2
a. F
b. T
c. F
d. T
e. F
f. F

Page 104 Exercise 3
Pituitary – Human growth hormone
Thyroid-stimulating hormone
Follicle-stimulating hormone
Luteinizing hormone
Prolactin
Adrenocorticotropic hormone
Melanocyte-stimulating hormone
Oxytocin
Antidiuretic hormone
Pineal – Melatonin
Thyroid – Thyroid hormone, Calcitonin
Parathyroids – Parathormone
Thymus – Thymosin
Pancreas – Glucagon, Insulin, Somatostatin
Adrenal cortex – Mineralcorticoids, Glucocorticoids, Sex hormones
Adrenal medulla – Adrenaline, Noradrenaline
Ovaries – Oestrogens, Progesterone
Testes – Testosterone

Exercise 4
1. hyposecretion
2. Diabetes mellitus
3. calcium
4. Seasonal Affective Disorder
5. acromegaly, gigantism
6. antidiuretic
7. goitre
8. hypothyroidism
9. Graves'
10. thyroid

Page 105 Vocabulary test
a. The normal, ordered death and removal of cells
b. The study of the endocrine glands and the hormones they secrete
c. Sex organs that produce mature sex cells
d. Over or excessive secretion
e. Under secretion

Multiple choice questions
1. b
2. a
3. d
4. a
5. b
6. c
7. c

8. d
9. b
10. a
11. c
12. a
13. c
14. b
15. d

Chapter 8
Page 112 Exercise 1
1. d
2. a
3. e
4. f
5. g
6. c
7. b

Exercise 2
a. Nasal cavity
b. Larynx
c. Trachea
d. Bronchi
e. Right lung
f. Pharynx
g. Left lung

Page 113 Exercise 3
1. b and h
2. i and d
3. e
4. c
5. g
6. a
7. f

Exercise 4
a. arteries, deoxygenated, oxygenated, veins
b. bronchial, pulmonary, bronchial, vena cava

Page 114 Exercise 5
a. T
b. F
c. T
d. T
e. F
f. F
g. T

Exercise 6
1. Asthma
2. Emphysema
3. Hyperventilation

4. Laryngitis
5. Rhinitis
6. Tuberculosis

Page 115 Vocabulary test
1. c
2. d
3. a
4. b

Multiple choice questions
1. a
2. c
3. b
4. c
5. d
6. a
7. c
8. b
9. a
10. d
11. d
12. b
13. c
14. a
15. a

Chapter 9
Page 126 Exercise 1
Across
3. Spleen
4. Solvent
6. Phagocytosis
8. Lymphocyte
9. Leucocyte
14. Electrolytes
15. Fibrinogen
17. Plasma

Down
1. Waste
2. Haemoglobin
5. Basophil
7. Blood
10. Erythrocyte
11. Haemostasis
12. Neutrophil
13. Hormone
16. Oxygen

Exercise 2
a. atrium, ventricle
b. right, deoxygenated, lungs
c. left, oxygenated, body
d. backwards, atrioventricular, semilunar

Exercise 3
a. Oxygenated
b. Heart
c. Heart
d. Heart tissue
e. Deoxygenated
f. Lungs
g. Most of the body superior to the diaphragm
h. Heart
i. Deoxygenated
j. Body inferior to the diaphragm
k. Deoxygenated
l. Oxygenated
m. Lungs

Page 128 Exercise 4
a. Pulmonary veins
b. Aorta and branches
c. Left atrium
d. Left ventricle
e. Right ventricle
f. Right atrium
g. Vena cava
h. Pulmonary arteries

Exercise 5
a. Sinoatrial node
b. Systole
c. Cardiac cycle
d. Autorhythmic cells
e. Arterioles
f. Diastole

Page 129 Exercise 6
a. Dave's blood pressure is high. Normal blood pressure is approximately 120/80mm Hg
b. The first figure, 160, is the systolic pressure and shows the pressure of the blood during contraction. The second figure, 100, is the diastolic pressure and shows the pressure of the blood during relaxation

Exercise 7
1. b
2. f
3. c
4. a
5. e
6. d
7. g
8. h

Page 130 Exercise 8
1. F
2. T
3. F
4. F
5. T
6. T
7. F
8. F
9. T
10. F

Vocabulary test
a. Relaxation of the heart muscle during the cardiac cycle
b. Contraction of the heart muscle during the cardiac cycle
c. The constriction of blood vessels
d. The dilation of blood vessels

Page 131
Multiple choice questions
1. b
2. a
3. c
4. a
5. d
6. d
7. a
8. a
9. d
10. c
11. b
12. a
13. d
14. d
15. c

Chapter 10
Page 136 Exercise 1
a. Drains interstitial fluid
b. Transports dietary lipids
c. Protects against invasion

Exercise 2
a. F
b. T
c. T
d. F
e. T

Exercise 3
a. 4
b. 3
c. 1
d. 2

Page 137 Exercise 4
Across
Supratrochlear
Superficial parotid
Mastoid
Ileocolic
Cervical
Inguinal

Down
Submandibular
Submental
Axillary
Occipital
Iliac
Popliteal

Page 138 Exercise 5
Non-specific resistance to disease
Mechanical barriers, natural killer cells, fever, chemical barriers, phagocytes, inflammation

The immune response
T-cells, B-cells, immunological memory

Exercise 6
1. Lymphoedema
2. Leukaemia
3. Hodgkin's disease
4. AIDS
5. Glandular fever

Page 139
Vocabulary test
a. A specialised protein that is synthesised to destroy a specific antigen
b. Any substance that the body recognises as foreign
c. A scavenger cell that engulfs and destroys microbes
d. The spread of disease from its site of origin
e. A cell that is able to engulf and digest microbes

Multiple choice questions
1. b
2. c
3. a
4. b
5. d
6. a
7. b

8. c
9. d
10. d
11. a
12. d
13. c
14. d
15. b

Chapter 11
Page 146 Exercise 1
Across
4. Caecum
5. Chyme
6. Mastication
8. Colon
10. Defecation
12. Mouth
13. Parotid
16. Lipases
17. Bile
18. Alimentary canal
20. Dentes
21. Bolus

Down
1. Anus
2. Deglutition
3. Amylase
7. Duodenum
9. Peritoneum
11. Ingestion
12. Monosaccharides
13. Peristalsis
14. Liver
15. Oesophagus
19. Mucosa

Page 148 Exercise 2
Mouth – salivary amylase, lingual lipase
Stomach – gastric lipase, pepsin, rennin
Small intestine – pancreatic amylase, trypsin, pancreatic lipase, bile, intestinal juice, brush border enzymes
Large intestine – bacteria

Exercise 3
a. T
b. F
c. F
d. T
e. F

Page 149
Exercise 4
1. g
2. h
3. a
4. f
5. b
6. e
7. c
8. d

Page 150 Exercise 5
1. c
2. h
3. b
4. f
5. a
6. j
7. d
8. e
9. g
10. i

Vocabulary test
a. The uptake of digested nutrients into the bloodstream and lymphatic system
b. A substance that alters the rate of a chemical reaction without itself being changed by the reaction
c. The process by which large molecules of food are broken down into smaller molecules that can enter cells
d. The substance on which an enzyme acts

Page 151
Multiple choice questions
1. b
2. d
3. a
4. c
5. a
6. b
7. c
8. c
9. d
10. a
11. b
12. d
13. b
14. c
15. b

Chapter 12
Page 158 Exercise 1
1. f
2. d
3. e
4. a
5. b
6. c

Exercise 2
a. blood, urine
b. adrenal/suprarenal
c. cortex, medulla, pelvis
d. nephrons

Page 159 Exercise 3
1. a
2. f
3. d
4. h
5. g
6. i
7. e
8. b
9. c

Page 160 Exercise 4
a. Afferent arteriole
b. Efferent arteriole
c. Proximal convoluted tubule
d. Distal convoluted tubule
e. Collecting duct
f. Medullary loop (Loop of Henle)
g. Glomerular capsule
h. Glomerulus
i. Branch of renal vein
j. Branch of renal artery

Exercise 5
Haemoglobin, pus, glucose, proteins, red blood cells, bile pigments

Page 161 Exercise 6
a. T
b. F
c. T
d. F
e. T

Exercise 7
1. e
2. a
3. d
4. b
5. c

Vocabulary test
a. A substance that increases urine production
b. A charged particle (ion) that conducts an electrical current in an aqueous solution
c. Urination

Page 162
Multiple choice questions
1. b
2. a
3. c
4. d
5. b
6. a
7. c
8. d
9. b
10. a
11. c
12. b
13. a
14. c
15. b

Chapter 13
Page 169 Exercise 1
a. Meiosis is reproductive cell division and results in sperm and egg cells being produced. Together these cells can form a new organism
b. Mitosis is somatic cell division and it occurs when the body needs to replace dead and injured cells or produce new cells for growth

Exercise 2
a. 4
b. 2
c. No
d. Yes
e. 23
f. 46

Page 170 Exercise 3
Male
Scrotum, penis, testes, epididymis, bulbourethral glands, spermatic cord, prostate, seminal vesicles, vas deferens

Female
Ovaries, vulva, vagina, Fallopian tubes, mammary glands, uterus

Exercise 4
a. F
b. T
c. F
d. F
e. T

Page 171 Exercise 5
a. Menstruation
b. Oocyte
c. Fallopian tubes
d. Endometrium
e. Corpus luteum
f. Graafian follicle

Exercise 6
The correct description is **B**

Page 172 Exercise 7
1. Prostatitis
2. Ectopic pregnancy
3. Mastitis
4. Amenorrhea
5. Polycystic ovary syndrome (Stein-Leventhal syndrome)

Vocabulary test
a. Fertilisation
b. Haploid cell
c. Lactation
d. Oogenesis
e. Semen (seminal fluid)

Page 173
Multiple choice questions
1. b
2. c
3. d
4. a
5. c
6. d
7. d
8. c
9. b
10. a
11. d
12. c
13. a
14. b
15. b

Answer key for mock exam papers

Paper 1

1.a 2.b 3.d 4.b 5.d 6.c 7.c 8.d 9.c 10.b 11.c 12.b 13.a 14.c 15.a 16.b 17.b 18.a 19.d 20.a 21.d 22.c 23.a 24.c 25.d

Paper 2

1.b 2.c 3.b 4.a 5.b 6.c 7.d 8.d 9.b 10.d 11.a 12.b 13.b 14.a 15.d 16.c 17.a 18.c 19.c 20.c 21.b 22.a 23.d 24.c 25.c

Paper 3

1.d 2.b 3.a 4.c 5.b 6.d 7.b 8.a 9.c 10.d 11.b 12.a 13.d 14.c 15.c 16.b 17.a 18.b 19.c 20.b 21.b 22.a 23.d 24.a 25.d

Paper 4

1.c 2.b 3.a 4.c 5.b 6.c 7.b 8.a 9.a 10.d 11.b 12.b 13.a 14.d 15.c 16.c 17.d 18.a 19.c 20.a 21.c 22.b 23.a 24.b 25.d

Paper 5

1.c 2.d 3.c 4.b 5.d 6.c 7.b 8.c 9.d 10.b 11.a 12.c 13.c 14.d 15.c 16.d 17.b 18.c 19.b 20.a 21.c 22.d 23.d 24.b 25.c 26.d 27.c 28.a 29.b 30.c 31.c 32.a 33.a 34.d 35.b 36.d 37.c 38.c 39.a 40.b 41.c 42.c 43.d 44.b 45.b 46.d 47.c 48.c 49.a 50.b

Paper 6

1.d 2.a 3.d 4.c 5.c 6.d 7.c 8.a 9.a 10.b 11.d 12.c 13.c 14.b 15.d 16.b 17.a 18.d 19.b 20.c 21.d 22.c 23.a 24.b 25.b 26.d 27.c 28.d 29.d 30.a 31.b 32.d 33.c 34.a 35.a 36.c 37.b 38.c 39.a 40.d 41.d 42.d 43.c 44.c 45.c 46.b 47.c 48.a 49.d 50.d

Paper 7

1.c 2.a 3.b 4.a 5.d 6.c 7.d 8.b 9.d 10.d 11.c 12.a 13.c 14.d 15.c 16.a 17.c 18.d 19.c 20.c 21.c 22.b 23.d 24.b 25.c 26.d 27.b 28.d 29.c 30.d 31.b 32.b 33.d 34.d 35.a 36.c 37.a 38.b 39.c 40.d 41.a 42.c 43.b 44.d 45.d 46.a 47.d 48.c 49.b 50.d

Paper 8

1.b 2.a 3.c 4.a 5.b 6.a 7.b 8.d 9.d 10.c 11.b 12.b 13.d 14.c 15.b 16.c 17.d 18.a 19.d 20.a 21.a 22.d 23.a 24.b 25.a 26.b 27.d 28.b 29.a 30.b 31.c 32.c 33.a 34.b 35.d 36.d 37.b 38.a 39.b 40.c 41.a 42.b 43.c 44.c 45.b 46.c 47.b 48.b 49.c 50.b

Paper 9

1.a 2.b 3.c 4.a 5.d 6.c 7.d 8.a 9.c 10.b 11.d 12.b 13.c 14.d 15.d 16.b 17.a 18.c 19.b 20.c 21.b 22.b 23.a 24.d 25.d 26.b 27.c 28.b 29.a 30.b 31.a 32.c 33.d 34.d 35.a 36.c 37.b 38.b 39.a 40.d 41.c 42.d 43.a 44.b 45.c 46.d 47.a 48.d 49.b 50.c

Paper 10

1.d 2.b 3.d 4.c 5.b 6.a 7.c 8.b 9.d 10.d 11.a 12.b 13.d 14.d 15.a 16.c 17.b 18.b 19.c 20.a 21.b 22.c 23.a 24.b 25.a 26.b 27.c 28.d 29.b 30.c 31.a 32.c 33.a 34.b 35.c 36.c 37.d 38.c 39.b 40.c 41.c 42.b 43.d 44.d 45.a 46.b 47.a 48.d 49.c 50.a

Paper 11

1.d 2.a 3.a 4.c 5.b 6.b 7.c 8.b 9.d 10.d 11.c 12.b 13.d 14.b 15.a 16.d 17.d 18.c 19.d 20.b 21.a 22.c 23.a 24.c 25.d 26.a 27.a 28.c 29.b 30.d 31.b 32.c 33.b 34.a 35.d 36.b 37.b 38.d 39.c 40.b 41.a 42.d 43.b 44.d 45.a 46.a 47.c 48.b 49.d 50.a

Paper 12

1.d 2.c 3.a 4.a 5.b 6.b 7.c 8.b 9.d 10.b 11.a 12.d 13.b 14.c 15.d 16.a 17.a 18.b 19.c 20.b 21.c 22.a 23.b 24.d 25.b 26.b 27.a 28.d 29.d 30.c 31.d 32.a 33.c 34.b 35.d 36.c 37.b 38.d 39.b 40.a 41.c 42.c 43.a 44.d 45.d 46.a 47.c 48.d 49.b 50.c

Paper 13

1.a 2.b 3.c 4.a 5.c 6.c 7.d 8.c 9.b 10.a 11.c 12.b 13.c 14.d 15.a 16.d 17.c 18.a 19.c 20.d 21.d 22.c 23.b 24.b 25.c 26.a 27.d 28.b 29.c 30.a 31.a 32.b 33.d 34.c 35.d 36.c 37.a 38.b 39.b 40.c 41.b 42.d 43.a 44.c 45.a 46.a 47.c 48.d 49.b 50.a

Paper 14

1.a 2.b 3.d 4.b 5.c 6.c 7.b 8.c 9.a 10.c 11.a 12.c 13.a 14.b 15.d 16.c 17.c 18.d 19.a 20.c 21.a 22.c 23.c 24.d 25.b 26.c 27.a 28.d 29.b 30.d 31.d 32.c 33.b 34.a 35.c 36.d 37.b 38.c 39.d 40.a 41.b 42.c 43.b 44.c 45.b 46.a 47.c 48.a 49.b 50.d